Mathematical Biosciences Institute Lecture Series

The Mathematical Biosciences Institute (MBI) fosters innovation in the application of mathematical, statistical and computational methods in the resolution of significant problems in the biosciences, and encourages the development of new areas in the mathematical sciences motivated by important questions in the biosciences. To accomplish this mission, MBI holds many week-long research workshops each year, trains postdoctoral fellows, and sponsors a variety of educational programs.

The MBI lecture series are readable, up to date collections of authored volumes that are tutorial in nature and are inspired by annual programs at the MBI. The purpose is to provide curricular materials that illustrate the applications of the mathematical sciences to the life sciences. The collections are organized as independent volumes, each one suitable for use as a (two-week) module in standard graduate courses in the mathematical sciences and written in a style accessible to researchers, professionals, and graduate students in the mathematical and biological sciences. The MBI lectures can also serve as an introduction for researchers to recent and emerging subject areas in the mathematical biosciences.

Marty Golubitsky, Michael Reed
Mathematical Biosciences Institute

Mathematical Biosciences Institute Lecture Series
Volume 1: Stochastics in Biological Systems

Stochasticity is fundamental to biological systems. In some situations the system can be treated as a large number of similar agents interacting in a homogeneously mixing environment, and so the dynamics are well-captured by deterministic ordinary differential equations. However, in many situations, the system can be driven by a small number of agents or strongly influenced by an environment fluctuating in space and time. For example, fluctuations are critical in the early stages of an epidemic; a small number of molecules may determine the direction of cellular processes; changing climate may alter the balance among competing populations. Spatial models may be required when agents are distributed in space and interactions between agents are local. Systems can evolve to become more robust or co-evolve in response to competitive or host-pathogen interactions. Consequently, models must allow agents to change and interact in complex ways. Stochasticity increases the complexity of models in some ways, but may also simplify and smooth results in other ways.

Volume 1 provides a series of lectures by well-known international researchers based on the year on Stochastics in biological systems which took place at the MBI in 2011–2012.

Michael Reed, Richard Durrett
Editors

Mathematical Biosciences Institute Lecture Series
Volume 1: Stochastics in Biological Systems

David F. Anderson • Thomas G. Kurtz

Stochastic Analysis
of Biochemical Systems

 Springer

Mathematical Biosciences Institute
at The Ohio State University

David F. Anderson
Department of Mathematics
University of Wisconsin
Madison, WI, USA

Thomas G. Kurtz
Departments of Mathematics and Statistics
University of Wisconsin
Madison, WI, USA

ISSN 2364-2297   ISSN 2364-2300   (electronic)
Mathematical Biosciences Institute Lecture series
ISBN 978-3-319-16894-4   ISBN 978-3-319-16895-1   (eBook)
DOI 10.1007/978-3-319-16895-1

Library of Congress Control Number: 2015935945

Mathematics Subject Classification (2010): 60J27, 60J28, 60J80, 60F05, 60F17, 60G44, 65C05, 65C20, 80A30, 92C40, 92C42

Springer Cham Heidelberg New York Dordrecht London

Springer International Publishing AG Switzerland is part of Springer Science+Business Media (www. springer.com)

Preface

This book, as with others in the series, is intended to provide supplementary material for courses in probability or stochastic processes. The mathematical focus is on counting processes and continuous-time Markov chains, and the selection of material is motivated by the examples and applications drawn from chemical networks in systems biology. While the material is presented in a manner most suitable for students who have studied stochastic processes up to and including martingales in continuous time, much of the necessary background material is summarized in the Appendix. Our hope is that a student with a solid understanding of calculus, differential equations, and elementary probability, and who is well-motivated by the applications, will be able to follow the main material with occasional reference to the Appendix.

As a review of the references will indicate, this text includes much work done by the authors with a long list of collaborators, work that was supported by a variety of grants from the National Science Foundation, most recently DMS-11-06424 and DMS-13-18832. This collaboration and support is gratefully acknowledged.

Madison, WI, USA   David F. Anderson
December 2014   Thomas G. Kurtz

Contents

Chapter 1
Infinitesimal specification of continuous time Markov chains

1.1 Poisson and general counting processes

The basic building blocks of the models we will consider are counting processes, that is, processes N such that $N(t)$ is the number of times that a particular phenomenon has been observed by time t. We assume that these observations occur one at a time, so we have the following definition.

Definition 1.1. N is a *counting process* if $N(0) = 0$ and N is constant except for jumps of $+1$.

If N is a counting process and $t < s$, then $N(s) - N(t)$ is the number of observations in the time interval $(t, s]$. The simplest counting process is a Poisson process.

Definition 1.2. A counting process is a *Poisson process* if it satisfies the following conditions:

1) Numbers of observations in disjoint time intervals are independent random variables, i.e., if $t_0 < t_1 < \cdots < t_m$, then $N(t_k) - N(t_{k-1})$, $k = 1, \ldots, m$ are independent random variables.
2) The distribution of $N(t + a) - N(t)$ does not depend on t.

Theorem 1.3. *If N is a Poisson process, then there is a constant $\lambda > 0$ such that, for $t < s$, $N(s) - N(t)$ is Poisson distributed with parameter $\lambda(s - t)$, that is,*

$$P\{N(s) - N(t) = k\} = \frac{(\lambda(s-t))^k}{k!} e^{-\lambda(s-t)}.$$

© Springer International Publishing Switzerland 2015
D.F. Anderson, T.G. Kurtz, *Stochastic Analysis of Biochemical Systems*,
Mathematical Biosciences Institute Lecture Series 1,
DOI 10.1007/978-3-319-16895-1_1

We will refer to N as a *unit* Poisson process if $\lambda = 1$.

Proof. For an integer $n > 0$, let $N_n(t)$ be the number of time intervals $(\frac{k}{n}, \frac{k+1}{n}]$, $k = 0, \ldots, [nt]$ that contain at least one observation. Then $N_n(t)$ is binomially distributed with parameters $[nt]$ and $p_n = P\{N(\frac{1}{n}) > 0\}$. Consequently,

$$P\{N(1) = 0\} = P\{N_n(1) = 0\} = (1 - p_n)^n.$$

Taking logs and noting that the left-hand side does not depend upon n, we may conclude that $np_n \to \lambda \equiv -\log P\{N(1) = 0\}$, as $n \to \infty$. The rest follows by the standard Poisson approximation of the binomial. $\qquad\square$

Let N be a Poisson process, and let S_k be the time of the kth observation, that is, the kth jump of N. Then

$$P\{S_k \le t\} = P\{N(t) \ge k\} = 1 - \sum_{i=0}^{k-1} \frac{(\lambda t)^i}{i!} e^{-\lambda t}, \quad t \ge 0.$$

Differentiating to obtain the probability density function gives

$$f_{S_k}(t) = \begin{cases} \frac{1}{(k-1)!} \lambda (\lambda t)^{k-1} e^{-\lambda t} & t \ge 0 \\ 0 & t < 0, \end{cases}$$

and we see that S_k is Γ-distributed. The proof of Theorem 1.3 can be refined to give the following:

Theorem 1.4. *Let $T_1 = S_1$ and for $k > 1$, $T_k = S_k - S_{k-1}$. Then $T_1, T_2, \ldots$ are independent and exponentially distributed with parameter λ.*

Watanabe's characterization of Poisson processes [47] is essential for our approach to more general counting processes. $\{\mathscr{F}_t^N\}$ will denote the filtration generated by N, that is, $\mathscr{F}_t^N = \sigma(N(s), s \le t)$ is the information obtained by observing N on the time interval $[0, t]$.

Theorem 1.5 (Watanabe). *If N is a Poisson process with parameter λ, then $N(t) - \lambda t$ is a martingale. Conversely, if N is a counting process and $N(t) - \lambda t$ is a martingale, then N is a Poisson process with parameter λ.*

Proof. The fact that for a Poisson process $N(t) - \lambda t$ is a martingale, that is,

$$E[N(t+s) - \lambda(t+s)|\mathscr{F}_t^N] = N(t) - \lambda t, \tag{1.1}$$

follows from the independent increments property and the fact that $E[N(t)] = \lambda t$.

Assuming the conditions of the converse, let $\{s_k\}$ be a partition of $(t, t+r]$. Then

$$E[e^{i\theta(N(t+r)-N(t))}|\mathscr{F}_t^N]$$

$$= 1 + \sum_{k=0}^{n-1} E[(e^{i\theta(N(s_{k+1})-N(s_k))} - 1)e^{i\theta(N(s_k)-N(t))}|\mathscr{F}_t^N]$$

$$= 1 + \sum_{k=0}^{n-1} E\Big[\Big(e^{i\theta(N(s_{k+1})-N(s_k))} - 1 - (e^{i\theta} - 1)(N(s_{k+1}) - N(s_k))\Big)$$

$$e^{i\theta(N(s_k)-N(t))}|\mathscr{F}_t^N\Big]$$

$$+ \sum_{k=0}^{n-1} \lambda(s_{k+1} - s_k)(e^{i\theta} - 1)E[e^{i\theta(N(s_k)-N(t))}|\mathscr{F}_t^N],$$

where the second equality holds by the martingale assumption which implies

$$E[N(s_{k+1}) - N(s_k)|\mathscr{F}_{s_k}^N] = \lambda(s_{k+1} - s_k).$$

We claim that as $\max_k(s_{k+1} - s_k) \to 0$, the first conditional expectation on the right-hand side of the final equality in the equation array above converges to zero. To see this, simply move the sum inside the conditional expectation and note that

$$\left|\sum_{k=0}^{n-1} \Big(e^{i\theta(N(s_{k+1})-N(s_k))} - 1 - (e^{i\theta} - 1)(N(s_{k+1}) - N(s_k))\Big)e^{i\theta(N(s_k)-N(t))}\right| \leq 4N(t)$$

and that the expression inside the first parentheses is zero if $N(s_{k+1}) - N(s_k)$ is 0 or 1. Thus, as $\max_k(s_{k+1} - s_k) \to 0$ the conditional expectation converges to zero by the dominated convergence theorem for conditional expectations (see Appendix A.1.2). Consequently, letting $\max_k(s_{k+1} - s_k) \to 0$, we have

$$E[e^{i\theta(N(t+r)-N(t))}|\mathscr{F}_t^N] = 1 + \lambda(e^{i\theta} - 1)\int_0^r E[e^{i\theta(N(t+s)-N(t))}|\mathscr{F}_t^N]ds$$

and $E[e^{i\theta(N(t+r)-N(t))}|\mathscr{F}_t^N] = e^{\lambda(e^{i\theta}-1)r}$, which implies both the independent increments property and the fact that $N(t+r) - N(t)$ is Poisson distributed with parameter λr. (See Section A.5.)   □

The filtration of interest may involve more information than just observations of N, and the calculations in the previous proof are still valid for any filtration $\{\mathscr{F}_t\}$ satisfying (1.1) with $\mathscr{F}_t^N$ replaced by $\mathscr{F}_t$, that is, M defined by $M(t) = N(t) - \lambda t$ is a martingale with respect to $\{\mathscr{F}_t\}$. In particular, N will be compatible with $\{\mathscr{F}_t\}$ in the following sense.

Definition 1.6. A Poisson process N is *compatible* with a filtration $\{\mathscr{F}_t\}$, if N is $\{\mathscr{F}_t\}$-adapted and $N(t + \cdot) - N(t)$ is independent of $\mathscr{F}_t$ for every $t \geq 0$.

Lemma 1.7. *Let N be a Poisson process with parameter $\lambda > 0$ that is compatible with $\{\mathscr{F}_t\}$, and let τ be a $\{\mathscr{F}_t\}$-stopping time such that $\tau < \infty$ a.s. Define*

$N_\tau(t) = N(\tau + t) - N(\tau)$. *Then N_τ is a Poisson process that is independent of $\mathscr{F}_\tau$ and compatible with $\{\mathscr{F}_{\tau+t}\}$.*

Proof. Let $M(t) = N(t) - \lambda t$. By the optional sampling theorem,

$$E[M((\tau + t + r) \wedge T)|\mathscr{F}_{\tau+t}] = M((\tau + t) \wedge T),$$

so

$$E[N((\tau + t + r) \wedge T) - N((\tau + t) \wedge T)|\mathscr{F}_{\tau+t}] = \lambda((\tau + t + r) \wedge T - (\tau + t) \wedge T).$$

Letting $T \to \infty$, by the monotone convergence theorem,

$$E[N(\tau + t + r) - N(\tau + t)|\mathscr{F}_{\tau+t}] = \lambda r,$$

which shows that M_τ defined by $M_\tau(t) = N_\tau(t) - t$ is a $\{\mathscr{F}_{\tau+t}\}$-martingale. By Theorem 1.5, we may conclude that N_τ is a Poisson process, and by the proof of Theorem 1.5, we further conclude that $N_\tau(t + \cdot) - N_\tau(t)$ is independent of $\mathscr{F}_{\tau+t}$. $\square$

If N is a Poisson process with parameter λ and N is compatible with $\{\mathscr{F}_t\}$, then

$$P\{N(t + \Delta t) > N(t)|\mathscr{F}_t\} = 1 - e^{-\lambda \Delta t} \approx \lambda \Delta t. \tag{1.2}$$

The parameter λ is referred to as the intensity for the Poisson process, and (1.2) suggests how to extend the notion of intensity to more general counting processes. Note that if Y is a unit Poisson process, then for $\lambda > 0$, $N(t) = Y(\lambda t)$ defines a Poisson process with intensity λ.

At least intuitively, a nonnegative, $\{\mathscr{F}_t\}$-adapted stochastic process $\lambda(\cdot)$ is an $\{\mathscr{F}_t\}$-*intensity* for N if

$$P\{N(t + \Delta t) > N(t)|\mathscr{F}_t\} \approx E[\int_t^{t+\Delta t} \lambda(s)ds|\mathscr{F}_t] \approx \lambda(t)\Delta t.$$

The following makes this definition precise. (See [33].)

Definition 1.8. Let N be a counting process adapted to $\{\mathscr{F}_t\}$, and let S_n be the nth jump time of N. A nonnegative $\{\mathscr{F}_t\}$-adapted stochastic process λ is an $\{\mathscr{F}_t\}$-*intensity* for N if and only if for each $n - 1, 2, \ldots,$

$$N(t \wedge S_n) - \int_0^{t \wedge S_n} \lambda(s)ds$$

is a $\{\mathscr{F}_t\}$-martingale.

If $\lim_{n \to \infty} S_n = \infty$, this requirement is equivalent to the requirement that

$$N(t) - \int_0^t \lambda(s)ds$$

be a local $\{\mathscr{F}_t\}$-martingale.

1.2 Modeling with intensities

In general the intensity for a counting process at time t may depend on the behavior of the counting process prior to time t, and it may also depend on other stochastic inputs. Consequently, if we want to specify a counting process model by specifying its intensity, we need to specify a nonnegative function that depends on the past of the counting process and perhaps other stochastic inputs.

With this idea in mind, let Z be a stochastic process that models "external noise" or the "environment." To avoid measurability issues, assume that Z is *cadlag* (that is, Z is right continuous with left limits at all $t > 0$) and E-valued for some complete, separable metric space (E, r). $D_E[0, \infty)$ will denote the space of cadlag, E-valued paths, and $D^c[0, \infty)$ will denote the space of counting paths (zero at time zero and constant except for jumps of $+1$) that are finite for all time, while $D^{c, \infty}[0, \infty)$ will include paths x that hit infinity in finite time with $x(t) = \infty$ beyond that time.

The Borel σ-algebra for $D_E[0, \infty)$ is the same for all of the standard topologies on $D_E[0, \infty)$ and is simply the σ-algebra generated by the evaluation functions $\pi_t : x \in D_E[0, \infty) \to x(t) \in E$.

Our model intensity will satisfy the following:

Condition 1.9.
$$\lambda : [0, \infty) \times D_E[0, \infty) \times D^c[0, \infty) \to [0, \infty)$$

is measurable and satisfies $\lambda(t, z, v) = \lambda(t, z^t, v^t)$, *where* $z^t(s) = z(s \wedge t)$ *and* $v^t(s) = v(s \wedge t)$ *(λ is* nonanticipating*), and*

$$\int_0^t \lambda(s, z, v) ds < \infty$$

for all $t \geq 0$, $z \in D_E[0, \infty)$ *and* $v \in D^c[0, \infty)$.

Let $Y = \{Y(u), u \geq 0\}$ be a unit Poisson process that is $\{\mathscr{G}_u\}$-compatible and assume that $Z(s)$ is $\mathscr{G}_0$-measurable for every $s \geq 0$. (In particular, Z is independent of Y.) Consider

$$N(t) = Y\left(\int_0^t \lambda(s, Z, N) ds\right). \tag{1.3}$$

Theorem 1.10. *There exists a unique solution of (1.3) up to* $S_\infty \equiv \lim_{n \to \infty} S_n$, *where* S_n *is the nth jump time of* N. *Further,* $\tau(t) = \int_0^t \lambda(s, Z, N) ds$ *is a* $\{\mathscr{G}_u\}$-*stopping time, and for each* $n = 1, 2, \ldots,$

$$N(t \wedge S_n) - \int_0^{t \wedge S_n} \lambda(s, Z, N) ds \tag{1.4}$$

is a $\{\mathscr{G}_{\tau(t)}\}$-*martingale. (In Definition 1.8,* $\mathscr{F}_t = \mathscr{G}_{\tau(t)}$.)

Proof. Existence and uniqueness follows by solving from one jump to the next.

Next, let $Y^r(u) = Y(r \wedge u)$ and let

$$N^r(t) = Y^r \left(\int_0^t \lambda(s, Z, N^r) ds \right).$$

If $\tau(t) = \int_0^t \lambda(s, Z, N) ds \leq r$, then $N^r(t) = N(t)$. Consequently,

$$\{\tau(t) \leq r\} = \left\{ \int_0^t \lambda(s, Z, N^r) ds \leq r \right\} \in \mathcal{G}_r,$$

with a similar statement for $\{\tau(t \wedge S_n) \leq r\}$. Since $M(u) = Y(u) - u$ is a $\{\mathcal{G}_u\}$-martingale, by the optional sampling theorem, for $T > 0$,

$$E[M(\tau((t+v) \wedge S_n) \wedge T) | \mathcal{G}_{\tau(t)}] = M(\tau((t+v) \wedge S_n)) \wedge \tau(t) \wedge T) = M(\tau(t \wedge S_n) \wedge T).$$

Letting $T \to \infty$, we see that (1.4) is a martingale. □

The stochastic equation (1.3) gives one way of characterizing a counting process modeled by specifying its intensity. A second common (and equivalent) approach is suggested by Watanabe's theorem, Theorem 1.5, and the martingale properties given in Theorem 1.10, that is, we can require the counting process to be a solution of a martingale problem (see [33]).

Definition 1.11. Let Z be a cadlag, E-valued stochastic process, and let λ satisfy Condition 1.9. A counting process N is a solution of the *martingale problem* for (λ, Z) if for each $n = 1, 2, \ldots$,

$$N(t \wedge S_n) - \int_0^{t \wedge S_n} \lambda(s, Z, N) ds$$

is a martingale with respect to the filtration

$$\mathcal{F}_t = \sigma(N(s), Z(r) : s \leq t, r \geq 0).$$

Clearly, the solution of (1.3) gives a solution of the martingale problem. In fact, essentially every solution of the martingale problem can be obtained as a solution of (1.3). The following theorem is a consequence of Watanabe's theorem.

Theorem 1.12. *If N is a solution of the martingale problem for (λ, Z), then N has the same distribution as the solution of the stochastic equation (1.3).*

Proof. First, suppose $\int_0^\infty \lambda(s, Z, N) ds = \infty$ a.s. Let $\gamma(u)$ satisfy

$$\gamma(u) = \inf \left\{ t : \int_0^t \lambda(s, Z, N) ds \geq u \right\}.$$

Then, since $\gamma(u+v) \geq \gamma(u)$,

$$E[N(\gamma(u+v) \wedge S_n \wedge T) - \int_0^{\gamma(u+v) \wedge S_n \wedge T} \lambda(s,Z,N)ds | \mathscr{F}_{\gamma(u)}]$$

$$= N(\gamma(u) \wedge S_n \wedge T) - \int_0^{\gamma(u) \wedge S_n \wedge T} \lambda(s,Z,N)ds.$$

A monotone convergence argument lets us send T and n to infinity. We then have

$$E[N(\gamma(u+v)) - (u+v) | \mathscr{F}_{\gamma(u)}] = N(\gamma(u)) - u,$$

so $Y(u) = N(\gamma(u))$ is a Poisson process. Since $\gamma(\tau(t)) = t$ we have that

$$N(t) = N(\gamma(\tau(t))) = Y(\tau(t)) = Y(\int_0^t \lambda(s,Z,N)ds),$$

and (1.3) is satisfied for this choice of Y. Finally, since $\mathscr{F}_0 \supset \sigma(Z(s), s \geq 0)$, Y is independent of Z.

If $\int_0^\infty \lambda(s,Z,N)ds < \infty$ with positive probability, then (perhaps on a larger sample space) let Y^* be a unit Poisson process that is independent of $\mathscr{F}_t$ for all $t \geq 0$ and consider $N^\varepsilon(t) = N(t) + Y^*(\varepsilon t)$. Then N^ε is a counting process with intensity $\lambda(t,Z,N) + \varepsilon$, and since $\int_0^\infty (\lambda(s,Z,N) + \varepsilon)ds = \infty$, we may follow the steps above to conclude that there is a Y^ε for which

$$N^\varepsilon(t) = Y^\varepsilon(\int_0^t (\lambda(s,Z,N) + \varepsilon)ds).$$

As $\varepsilon \to 0$, Y^ε converges to

$$Y(u) = \begin{cases} N(\gamma(u)) & u < \tau(\infty) \\ N(\infty) + Y^*(u - \tau(\infty)) & u \geq \tau(\infty) \end{cases}$$

(except at points of discontinuity), and N satisfies (1.3) for this choice of Y. □

1.3 Multivariate counting processes

The models we wish to specify often involve more than one counting process, so we want to be able to specify multiple interdependent intensities. Let $D_d^c[0,\infty) = D^c[0,\infty)^d$ be the collection of d-dimensional counting paths.

Condition 1.13. *For $k = 1,\dots,d$,*

$$\lambda_k : [0,\infty) \times D_E[0,\infty) \times D_d^c[0,\infty) \to [0,\infty)$$

is measurable and nonanticipating with

$$\int_0^t \sum_k \lambda_k(s,z,v)\,ds < \infty, \quad z \in D_E[0,\infty), v \in D_d^c[0,\infty).$$

Let Z be cadlag and E-valued and independent of independent Poisson processes $Y_1,\dots,Y_d$. Consider the system of equations

$$N_k(t) = Y_k\Big(\int_0^t \lambda_k(s,Z,N)\,ds\Big), \quad k=1,\dots,d, \tag{1.5}$$

where $N = (N_1,\dots,N_d)$. Setting $S_n = \inf\{t : \sum_k N_k(t) \geq n\}$ and $S_\infty = \lim_{n\to\infty} S_n$, existence and uniqueness holds (including for $d = \infty$) up to S_∞. For $t \geq S_\infty$, we define $N(t) = \lim_{n\to\infty} N(S_n)$.

For each k, it is not hard to see

$$M_k^n(t) = N_k(t \wedge S_n) - \int_0^{t \wedge S_n} \lambda_k(s,Z,N)\,ds$$

is a martingale, but it is not immediately clear that the M_k^n are martingales with respect to a common filtration. To construct the appropriate filtration, we need to understand filtrations and martingales indexed by directed sets.

Suppose $\mathscr{I}$ is a directed set (see Appendix A.6) with partial ordering $u \preceq v$. Let a collection of σ-algebras $\mathscr{G}_u \subset \mathscr{F}$, $\{\mathscr{G}_u, u \in \mathscr{I}\}$ be a filtration in the sense that $u \preceq v$ implies $\mathscr{G}_u \subset \mathscr{G}_v$. A stochastic process X indexed by $\mathscr{I}$ is a $\{\mathscr{G}_u\}$-martingale if and only if $E[|X(v)|] < \infty$ for all $v \in \mathscr{I}$ and for $u \preceq v$,

$$E[X(v)|\mathscr{G}_u] = X(u).$$

An $\mathscr{I}$-valued random variable τ is a stopping time if and only if $\{\tau \preceq u\} \in \mathscr{G}_u$, $u \in \mathscr{I}$, and as in the $\mathscr{I} = [0,\infty)$ case, define

$$\mathscr{G}_\tau = \{A \in \mathscr{F} : A \cap \{\tau \preceq u\} \in \mathscr{G}_u, u \in \mathscr{I}\}.$$

Lemma 1.14. *Let X be a martingale, and let τ_1 and τ_2 be stopping times assuming countably many values and satisfying $\tau_1 \preceq \tau_2$ a.s. If there exists a sequence $\{u_m\} \subset \mathscr{I}$ such that $\lim_{m\to\infty} P\{\tau_2 \preceq u_m\} = 1$, $\lim_{m\to\infty} E[|X(u_m)|\mathbf{1}_{\{\tau_2 \preceq u_m\}^c}] = 0$, and $E[|X(\tau_2)|] < \infty$, then*

$$E[X(\tau_2)|\mathscr{F}_{\tau_1}] = X(\tau_1). \tag{1.6}$$

Proof ([36]). Let $\Gamma \subset \mathscr{I}$ be countable and satisfy $P\{\tau_i \in \Gamma\} = 1$ and $\{u_m\} \subset \Gamma$. Define

$$\tau_i^m = \begin{cases} \tau_i & \text{on } \{\tau_i \preceq u_m\} \\ u_m & \text{on } \{\tau_i \preceq u_m\}^c. \end{cases}$$

Then τ_i^m is a stopping time, since

$$\begin{aligned}
\{\tau_i^m \preceq u\} &= (\{\tau_i^m \preceq u\} \cap \{\tau_i \preceq u_m\}) \cup (\{\tau_i^m \preceq u\} \cap \{\tau_i \preceq u_m\}^c) \\
&= (\{\tau_i \preceq u\} \cap \{\tau_i \preceq u_m\}) \cup (\{u_m \preceq u\} \cap \{\tau_i \preceq u_m\}^c) \\
&= (\cup_{\{v: v \in \Gamma, v \preceq u, v \preceq u_m\}} \{\tau_i \preceq v\}) \cup (\{u_m \preceq u\} \cap \{\tau_i \preceq u_m\}^c).
\end{aligned}$$

For $A \in \mathscr{G}_{\tau_1}$,

$$\begin{aligned}
\int_{A \cap \{\tau_1^m = t\}} X(\tau_2^m) dP &= \sum_{s \in \Gamma, s \preceq u_m} \int_{A \cap \{\tau_1^m = t\} \cap \{\tau_2^m = s\}} X(s) dP \\
&= \sum_{s \in \Gamma, s \preceq u_m} \int_{A \cap \{\tau_1^m = t\} \cap \{\tau_2^m = s\}} X(u_m) dP \\
&= \int_{A \cap \{\tau_1^m = t\}} X(u_m) dP \\
&= \int_{A \cap \{\tau_1^m = t\}} X(t) dP = \int_{A \cap \{\tau_1^m = t\}} X(\tau_1^m) dP.
\end{aligned}$$

Hence $E[X(\tau_2^m)|\mathscr{F}_{\tau_1}] = X(\tau_1^m)$, and letting $m \to \infty$, we have (1.6). $\qquad\square$

Let $\mathscr{I} = [0, \infty)^d$, where $\preceq$ denotes componentwise inequality. For $u \in \mathscr{I}$, set $\mathscr{G}_u = \sigma(Y_k(s_k) : s_k \leq u_k, k = 1, \ldots, d)$. Then

$$M_k(u) \equiv Y_k(u_k) - u_k$$

is a $\{\mathscr{G}_u\}$-martingale. For

$$N_k(t) = Y_k\left(\int_0^t \lambda_k(s, Z, N) ds\right),$$

define $\tau_k(t) = \int_0^t \lambda_k(s, Z, N) ds$ and $\tau(t) = (\tau_1(t), \ldots, \tau_d(t))$. By the same argument as in the proof of Theorem 1.10, $\tau(t)$ is a $\{\mathscr{G}_u\}$-stopping time, and similar arguments give the following.

Lemma 1.15. *Let $\mathscr{F}_t = \mathscr{G}_{\tau(t)}$. If σ is a $\{\mathscr{F}_t\}$-stopping time, then $\tau(\sigma)$ is a $\{\mathscr{G}_u\}$-stopping time.*

Lemma 1.16. *If τ is a $\{\mathscr{G}_u\}$-stopping time, then $\tau^{(n)}$ defined by*

$$\tau_k^{(n)} = \frac{[\tau_k 2^n] + 1}{2^n}$$

is a $\{\mathscr{G}_u\}$-stopping time.

Proof. For $u \in \mathscr{I}$,

$$\{\tau^{(n)} \preceq u\} = \cap_k \{\tau_k^{(n)} \leq u_k\} = \cap_k \{[\tau_k 2^n] + 1 \leq [u_k 2^n]\} = \cap_k \{\tau_k < \frac{[u_k 2^n]}{2^n}\}$$

which is in $\mathscr{G}_u$. $\qquad\square$

Note that $\tau_k^{(n)}$ decreases to τ_k.

Theorem 1.17. *Let Condition 1.13 hold. For $n = 1, 2, \ldots$, there exists a unique solution of (1.5) up to S_∞, $\tau_k(t) = \int_0^t \lambda_k(s, Z, N) ds$, $k = 1, \ldots, d$ defines a $\{\mathscr{G}_u\}$-stopping time, and*

$$N_k(t \wedge S_n) - \int_0^{t \wedge S_n} \lambda_k(s, Z, N) ds$$

is a $\{\mathscr{G}_{\tau(t)}\}$-martingale.

Proof. Approximating $\tau(t)$ by discrete stopping times as in Lemma 1.16, the result follows from the fact that $M_k(u)$ is a $\{\mathscr{G}_u\}$-martingale and Lemma 1.14. □

Definition 1.18. Let Z be a cadlag, E-valued stochastic process, and let $\lambda = (\lambda_1, \ldots, \lambda_d)$ satisfy Condition 1.13. A multivariate counting process N is a solution of the *martingale problem* for (λ, Z) if for each k,

$$N_k(t \wedge S_n) - \int_0^{t \wedge S_n} \lambda_k(s, Z, N) ds.$$

is a martingale with respect to the filtration

$$\mathscr{F}_t = \sigma(N(s), Z(r) : s \le t, r \ge 0).$$

For $t \ge S_\infty$, we define $N(t) = \lim_{n \to \infty} N(S_n)$.

Theorem 1.19. *Let Z be a cadlag, E-valued stochastic process, and let $\lambda = (\lambda_1, \ldots, \lambda_d)$ satisfy Condition 1.13. Then there exists a unique solution of the martingale problem for (λ, Z).*

Proof. Existence follows from the time-change equation, and uniqueness follows by Theorem 2 in [40] which essentially shows that every solution of the martingale problem can be written as a solution of the time-change equation. (See also Theorem 3.11 of [37].) □

1.4 Continuous time Markov chains

In most presentations, continuous time Markov chains are specified in terms of a q-matrix, where for $i \ne j$, q_{ij} gives

$$P\{X(t + \Delta t) = j | X(t) = i\} \approx q_{ij} \Delta t. \tag{1.7}$$

We will assume that X takes values in a subset $\mathbb{S}$ of a discrete lattice in $\mathbb{R}^d$ and specify intensities of jumps,

$$P\{X(t + \Delta t) - X(t) = \zeta_l | \mathscr{F}_t\} \approx \lambda_l(X(t)) \Delta t.$$

Of course, we could write

$$q_{X(t),X(t)+\zeta_l} = \lambda_l(X(t)),$$

but using our approach

$$X(t) = X(0) + \sum_l R_l(t)\zeta_l,$$

where R_l is a counting process counting the number of jumps of type ζ_l, and hence

$$X(t) = X(0) + \sum_l Y_l\left(\int_0^t \lambda_l(X(s))ds\right)\zeta_l. \tag{1.8}$$

For the moment, assume that there are only finitely many (λ_l, ζ_l) and that the λ_l are bounded. Then

$$\tilde{R}_l(t) = R_l(t) - \int_0^t \lambda_l(X(s))ds = Y_l\left(\int_0^t \lambda_l(X(s))ds\right) - \int_0^t \lambda_l(X(s))ds$$

is a martingale, and for a bounded function f,

$$f(X(t)) = f(X(0)) + \sum_l \int_0^t (f(X(s-) + \zeta_l) - f(X(s-)))dR_l(s)$$

$$= f(X(0)) + \sum_l \int_0^t (f(X(s-) + \zeta_l) - f(X(s-)))d\tilde{R}_l(s)$$

$$+ \sum_l \int_0^t \lambda_l(X(s))(f(X(s) + \zeta_l) - f(X(s)))ds.$$

For our purposes, we can define a *stochastic integral* for cadlag processes U and V by

$$\int_0^t U(s-)dV(s-) = \lim \sum U(t_i \wedge t)(V(t_{i+1} \wedge t) - V(t_i \wedge t)),$$

where $\{t_i\}$ is a partition of $[0,\infty)$ and the limit is taken as $\max_i(t_{i+1} - t_i) \to 0$, provided the limit exists in probability. If V is a $\{\mathscr{F}_t\}$-martingale and U is $\{\mathscr{F}_t\}$-adapted and bounded by a constant, then it is easy to check that each of the approximating sums is a martingale. Under our current assumptions, the limit will also be a martingale, or if we relax the boundedness assumptions on the λ_k, at least a local martingale.

Setting

$$Af(x) = \sum_l \lambda_l(x)(f(x + \zeta_l) - f(x)), \tag{1.9}$$

it follows that

$$f(X(t)) - f(X(0)) - \int_0^t Af(X(s))ds$$

$$= \sum_l \int_0^t (f(X(s-) + \zeta_l) - f(X(s-)))d\tilde{R}_l(s)$$

is a martingale. (See Appendix A.3.) We call A as defined in (1.9) the *generator* for the Markov chain.

Now we drop the finiteness and boundedness assumptions, but we do require $\sum_l \lambda_l(x) < \infty$ for each x. Define

$$\tau_K = \inf\{t : |X(t)| \geq K\}.$$

Then

$$E[\sum_l R_l(t \wedge \tau_K)] = E[\int_0^{t \wedge \tau_K} \sum_l \lambda_l(X(s))ds] \leq t \sup_{|x| \leq K} \sum_l \lambda_l(x)$$

and

$$f(X(t \wedge \tau_K)) - f(X(0)) - \int_0^{t \wedge \tau_K} Af(X(s))ds$$

is a martingale. Assume that if f has finite support, then $\lim_{|x| \to \infty} Af(x) = 0$. In addition, if $\lim_{K \to \infty} \tau_K \equiv \tau_\infty < \infty$, we define $X(t) = \infty$ for $t \geq \tau_\infty$. Then

$$f(X(t)) - f(X(0)) - \int_0^t Af(X(s))ds \qquad (1.10)$$

$$= \sum_l \int_0^t (f(X(s-) + \zeta_l) - f(X(s-)))d\tilde{R}_l(s)$$

is a martingale.

For a finite or countable index set $\mathscr{R}$, we assume the following:

Condition 1.20. • $\lambda_l(x) \geq 0$, $x \in \mathbb{S}$, $l \in \mathscr{R}$.
• $\zeta_l \in \mathbb{R}^d$, $l \in \mathscr{R}$, such that $x \in \mathbb{S}$ and $\lambda_l(x) > 0$ implies $x + \zeta_l \in \mathbb{S}$.
• For $x \in \mathbb{S}$, $\sum_{l \in \mathscr{R}} \lambda_l(x) < \infty$.
• For f with finite support in $\mathbb{S}$ and A defined by (1.9), $\lim_{|x| \to \infty} Af(x) = 0$.

The last condition is a modest restriction and is clearly satisfied for the chemical reaction network models we will consider in Chapter 2. We define $Af(\infty) = 0$.

Definition 1.21. Let A satisfy Condition 1.20. A right continuous, $\mathbb{S} \cup \{\infty\}$-valued stochastic process X is a solution of the *martingale problem* for A if there exists a filtration $\{\mathscr{F}_t\}$ such that for each f with finite support, (1.10) is a $\{\mathscr{F}_t\}$-martingale. If τ_∞ is finite with positive probability, then there may be more than one solution of the martingale problem. If in addition to the martingale requirements, we require that $X(t) = \infty$ for $t \geq \tau_\infty$, then we say that X is a *minimal solution* of the martingale problem.

Theorem 1.22. *Assume A satisfies Condition 1.20. Then the solution of*

$$X(t) = X(0) + \sum_{l \in \mathcal{R}} Y_l\left(\int_0^t \lambda_l(X(s))ds\right)\zeta_l$$

with $X(t) = \infty$ for $t \geq \tau_\infty$ is the unique minimal solution of the martingale problem for A.

The martingale property implies

$$E[f(X(t))] = E[f(X(0))] + \int_0^t E[Af(X(s))]ds,$$

and taking $f(x) = \mathbf{1}_{\{y\}}(x)$, we have

$$P\{X(t) = y\} = P\{X(0) = y\} + \int_0^t \left(\sum_l \lambda_l(y - \zeta_l)P\{X(s) = y - \zeta_l\}\right.$$
$$\left. - \sum_l \lambda_l(y)P\{X(s) = y\}\right)ds$$

giving the *Kolmogorov forward* or *master equation* for the one-dimensional distributions of X. In particular, defining $p_y(t) = P\{X(t) = y\}$ and $v_y = P\{X(0) = y\}$, $\{p_y\}$ satisfies the system of differential equations

$$\dot{p}_y(t) = \sum_l \lambda_l(y - \zeta_l)p_{y - \zeta_l}(t) - \sum_l \lambda_l(y)p_y(t), \tag{1.11}$$

with initial condition $p_y(0) = v_y$. We also require

$$p_y(t) \geq 0 \text{ and } \sum_y p_y(t) \leq 1, \quad t \geq 0. \tag{1.12}$$

Lemma 1.23. *Assume Condition 1.20. Let $\{v_y\}$ be a probability distribution on $\mathbb{S}$, and let $X(0)$ satisfy $P\{X(0) = y\} = v_y$. The solution of the system of differential equations (1.11) satisfying (1.12) and $p_y(0) = v_y$ is unique and the unique solution satisfies $\sum_y p_y(t) \equiv 1$ if and only if the solution of (1.8) satisfies $P\{\tau_\infty = \infty\} = 1$.*

Proof. Set $\overline{\lambda}(y) = \sum_l \lambda_l(y)$. The forward equation can be rewritten as

$$p_y(t) = v_y e^{-\overline{\lambda}(y)t} + \int_0^t e^{-\overline{\lambda}(y)(t-s)} \sum_l \lambda_l(y - \zeta_l)p_{y - \zeta_l}(s)ds. \tag{1.13}$$

Define $p_y^{(0)}(t) \equiv 0$ and iterate

$$p_y^{(k+1)}(t) = v_y e^{-\overline{\lambda}(y)t} + \int_0^t e^{-\overline{\lambda}(y)(t-s)} \sum_l \lambda_l(y - \zeta_l)p_{y - \zeta_l}^{(k)}(s)ds.$$

Let S_n be the nth jump time of the time-change equation (1.8). Note that

$$p_y^{(1)}(t) = P\{X(t) = y, S_1 > t\},$$

and in general

$$p_y^{(k)}(t) = P\{X(t) = y, S_k > t\}.$$

Observe that $\tau_\infty = \infty$ if and only if $S_\infty = \infty$, so if $P\{\tau_\infty = \infty\} = 1$,

$$1 = \sum_{y \in \mathbb{Z}^d} P\{X(t) = y\} = \lim_{k \to \infty} \sum_{y \in \mathbb{Z}^d} P\{X(t) = y, S_k > t\}.$$

Since every solution of (1.13) and (1.12) satisfies $p_y(t) \geq p_y^{(k)}(t)$, $\sum_y p_y^{(k)}(t) \to 1$ implies $p_y^{(k)}(t) \to p_y(t)$ giving uniqueness for (1.11).   □

Remark 1.24. Since $p_y^{(k)}(t)$ is monotone in k, it converges to the *minimal solution* (that is, the smallest solution) of the forward equation and that gives the distribution of the minimal solution of (1.8).

As the name suggests, a Markov chain is an example of a Markov process, that is, a stochastic process that has the following Markov property.

Definition 1.25. An $\mathbb{S} \cup \{\infty\}$-valued stochastic process X is *Markov* with respect to a filtration $\{\mathscr{F}_t\}$ if it is $\{\mathscr{F}_t\}$-adapted and

$$E[f(X(t+s))|\mathscr{F}_t] = E[f(X(t+s))|X(t)],$$

for all $s, t \geq 0$ and $f \in B(\mathbb{S} \cup \{\infty\})$. X is *strong Markov* with respect to $\{\mathscr{F}_t\}$ if for each finite $\{\mathscr{F}_t\}$-stopping time τ,

$$E[f(X(\tau+s))|\mathscr{F}_\tau] = E[f(X(\tau+s))|X(\tau)],$$

for all $s \geq 0$ and $f \in B(\mathbb{S} \cup \{\infty\})$.

Let X be the minimal solution of (1.8). Note that there exists a mapping

$$H : [0,\infty) \times \mathbb{S} \times D^{c,\infty}[0,\infty)^m \to D_{\mathbb{S} \cup \{\infty\}}[0,\infty)$$

such that

$$X(t) = H(t, X(0), \{Y_k\}).$$

Let $\tau_k(r) = \int_0^r \lambda_k(X(s)) ds$. For each $t \geq 0$,

$$Y_k^r(u) = Y_k(\tau_k(r) + u) - Y_k(\tau_k(r)), \quad k = 1, \ldots, m,$$

are independent Poisson processes independent of $\mathscr{F}_t = \mathscr{G}_{\tau(t)}$.
    Define $X^r(s) = X(r+s)$. Then

$$X^r(t) = X^r(0) + \sum_k Y_k^r \left(\int_0^t \lambda_k(X(s)) ds \right) \zeta_k$$

and

$$X^r(t) = H(t, X^r(0), \{Y_k^r\}),$$

that is, the future of X^r is a function of the present, $X^r(0)$, and inputs, $\{Y_k^r\}$, that are independent of the past. The same holds if r is replaced by a $\{\mathscr{F}_t\}$ stopping time γ. Consequently, X is strong Markov, and since H does not depend on r or γ,

$$P\{X(\gamma+t) = y | X(\gamma) = x\} = P\{X(t) = y | X(0) = x\}.$$

Problems

1.1. Prove Theorem 1.4.

1.2. Let $\lambda(n)$, $n = 0, 1, \ldots$ be positive and bounded (to avoid explosions), and let Y be a unit Poisson process. For

$$N(t) = Y\left(\int_0^t \lambda(N(s)) ds\right),$$

show that

$$P\{N(t + \Delta t) = N(t) | \mathscr{F}_t^N\} = e^{-\lambda(N(t))\Delta t},$$

where $\{\mathscr{F}_t^N\}$ is the filtration generated by N, so that

$$P\{N(t + \Delta t) > N(t) | \mathscr{F}_t^N\} = 1 - e^{-\lambda(N(t))\Delta t} \approx \lambda(N(t))\Delta t.$$

1.3. Let N_0 be a Poisson process with parameter λ, and let $\xi_1, \xi_2, \ldots$ be a sequence of Bernoulli trials with parameter p. Assume that the ξ_k are independent of N_0, and define

$$N_1(t) = \sum_{k=1}^{N_0(t)} \xi_k$$

and

$$N_2(t) = \sum_{k=1}^{N_0(t)} (1 - \xi_k).$$

a) What is the distribution of N_1? N_2?
b) Show that N_1 and N_2 are independent.
c) What is $P\{N_1(t) = k | N_0(t) = n\}$?
d) Let S_1 be the first jump time of N_0. For $k \geq 1$, find the conditional density of S_1 given that $N_0(t) = k$. (Hint: First calculate $P\{S_1 \leq s, N_0(t) = k\}$ for $s \leq t$.)

1.4. Let $Y_1, \ldots, Y_5$ be independent unit Poisson processes, and let $\lambda_1(n)$ and $\lambda_2(n)$ be positive, $n = 0, 1, \ldots$. Let

$$N_1(t) = Y_1\left(\int_0^t \lambda_1(N_1(s))ds\right)$$

$$N_2(t) = Y_2\left(\int_0^t \lambda_2(N_2(s))ds\right)$$

$$N_3(t) = Y_3\left(\int_0^t \lambda_1(N_3(s)) \wedge \lambda_2(N_4(s))ds\right)$$
$$+ Y_4\left(\int_0^t (\lambda_1(N_3(s)) - \lambda_1(N_3(s)) \wedge \lambda_2(N_4(s)))ds\right)$$

$$N_4(t) = Y_3\left(\int_0^t \lambda_1(N_3(s)) \wedge \lambda_2(N_4(s))ds\right)$$
$$+ Y_5\left(\int_0^t (\lambda_2(N_4(s)) - \lambda_1(N_3(s)) \wedge \lambda_2(N_4(s)))ds\right).$$

1. Show that N_3 has the same distribution as N_1 and N_4 has the same distribution as N_2.
2. Suppose that $\varepsilon > 0$ and $|\lambda_1(n) - \lambda_2(n)| \leq \varepsilon$. Let $\tau_1 = \inf\{t : N_1(t) \neq N_2(t)\}$ and $\tau_2 = \inf\{t : N_3(t) \neq N_4(t)\}$. Give a stochastic lower bound for τ_1 (that is, find a random variable σ_1 such that $P\{\tau_1 \geq t\} \geq P\{\sigma_1 \geq t\}$ for all t) and a stochastic lower bound for τ_2. In particular, show that τ_1 is stochastically less than τ_2.

1.5. Show that the solution of

$$N(t) = Y\left(\int_0^t (1 + N(s)^2)ds\right)$$

hits infinity in finite time.

More generally, show that if $\lambda(n) > 0$ satisfies

$$\sum_{n=0}^{\infty} \frac{1}{\lambda(n)} < \infty,$$

then with probability one, the solution of

$$N(t) = Y\left(\int_0^t \lambda(N(s))ds\right)$$

hits infinity in finite time.

1.6. Consider the infinite server queueing model

$$Q(t) = Y_1(\lambda t) - Y_2\left(\int_0^t \mu Q(s)ds\right).$$

Use martingale properties to compute $E[Q(t)]$ making sure to justify all the steps in the calculation.

1.7. Let $X_1(t)$ be a linear death process

$$X_1(t) = N - Y(\int_0^t \mu X_1(s)ds).$$

Let $\Delta_1, \ldots, \Delta_N$ be independent exponential random variables with parameter μ and define

$$X_2(t) = \sum_{i=1}^{N} 1_{\{\Delta_i > t\}}.$$

Show that X_1 and X_2 have the same distribution. (In other words, a linear death process models a population in which there are no births and each individual has an independent, exponentially distributed lifetime.)

1.8. Let $\{\Delta_k, k = 0, 1, \ldots\}$ be independent, unit exponential random variables. Define

$$p_k(t) = P\{\sum_{l=k+1}^{\infty} \frac{\Delta_l}{l^2} \le t < \sum_{l=k}^{\infty} \frac{\Delta_l}{l^2}\}.$$

Show that $\{p_k\}$ satisfies

$$\dot{p}_k(t) = (k+1)^2 p_{k+1}(t) - k^2 p_k(t),$$

that is, $\{p_k\}$ is a solution of an equation of the form (1.11), and show that the conclusion of Lemma 1.23 is in general false, if the requirement (1.12) is dropped.

Chapter 2
Models of biochemical reaction systems

2.1 The basic model

It is useful to understand that a biochemical reaction *system* consists of two parts: (i) a reaction *network*, and (ii) a choice of *dynamics*. The network is a static object that consists of a triple of sets:

(i) *species*, $\mathscr{S}$, which are the chemical components whose counts we wish to model dynamically,
(ii) *complexes*, $\mathscr{C}$, which are nonnegative linear combinations of the species that describe how the species can interact, and
(iii) *reactions*, $\mathscr{R}$, which describe how to convert one such complex to another.

For example, if in our system we have only three species, which we denote by A, B, and C, and the only transition type we allow is the merging of an A and a B molecule to form a C molecule, then we may depict this network by the directed graph

$$A + B \to C.$$

For this very simply model our network consists of species $\mathscr{S} = \{A, B, C\}$, complexes $\mathscr{C} = \{A + B, C\}$, and reactions $\mathscr{R} = \{A + B \to C\}$.

Before formally defining a reaction network, we provide a slightly less trivial example.

Example 2.1. Suppose there are two forms of a given protein: "active" and "inactive." Denote by A the active form of the protein and denote by B the inactive form. We suppose that there are only two types of transitions that can take place in the

The original version of this chapter was revised. An erratum to this chapter can be found at DOI 10.1007/978-3-319-16895-1_6

Electronic supplementary material The online version of this chapter (doi: 10.1007/978-3-319-16895-1_2) contains supplementary material, which is available to authorized users.

© Springer International Publishing Switzerland 2015
D.F. Anderson, T.G. Kurtz, *Stochastic Analysis of Biochemical Systems*,
Mathematical Biosciences Institute Lecture Series 1,
DOI 10.1007/978-3-319-16895-1_2

model: an active protein can become inactive, and an inactive protein can become active. However, we further suppose that an inactive protein B is required to catalyze the inactivation of an active protein A. That is, we suppose that the two possible reactions can be depicted in the following manner

$$A + B \rightarrow 2B, \tag{R1}$$

$$B \rightarrow A, \tag{R2}$$

where, for example, the reaction (R1) captures the idea that both an A and a B molecule are required for the deactivation of an A molecule and the result of such a reaction is a net gain of one molecule of B and a net loss of one molecule of A.

We again see that there are three sets of objects necessary to give a full description of the above network. First, we need a set of species, which in this case is just $\mathscr{S} = \{A, B\}$. We require a directed graph in which the vertices are linear combinations of the species. These linear combinations are the complexes, which for this model is the set $\mathscr{C} = \{A + B, \ 2B, \ B, \ A\}$. Finally, we associate the edges of the graph with the reactions, $\mathscr{R} = \{A + B \rightarrow 2B, \ B \rightarrow A\}$.   △

Definition 2.2. A chemical reaction network is a triple $\{\mathscr{S}, \mathscr{C}, \mathscr{R}\}$ where

(i) $\mathscr{S} = \{S_1, \ldots, S_n\}$ is the set of species,
(ii) $\mathscr{C}$ is the set of complexes, consisting of nonnegative linear combinations of the species,
(iii) $\mathscr{R} = \{y_k \rightarrow y'_k : y_k, y'_k \in \mathscr{C} \text{ and } y_k \neq y'_k\}$ is the set of reactions.

The notation we use throughout is to write the kth reaction as

$$\sum_i y_{ki} S_i \rightarrow \sum_i y'_{ki} S_i, \tag{2.1}$$

where the vectors $y_k, y'_k \in \mathbb{Z}^n_{\geq 0}$ are associated with the source and product complex, respectively. Note that we abuse notation slightly by writing $y_k \rightarrow y'_k$ as opposed to (2.1). We define $\zeta_k := y'_k - y_k \in \mathbb{Z}^n$ to be the *reaction vectors* of the network.

It is most common to forgo formally giving each of the three sets necessary for a reaction network, as it is easier to simply give the directed graph implied by the reaction network. For example, the network

$$S + E \rightleftharpoons C \rightarrow S + P, \qquad E \rightleftharpoons \emptyset, \tag{2.2}$$

corresponds to the reaction network with $\mathscr{S} = \{S, E, C, P\}, \mathscr{C} = \{S + E, \ C, \ S + P, \ E, \ \emptyset\}$, and $\mathscr{R} = \{S + E \rightarrow C, \ C \rightarrow S + E, \ C \rightarrow S + P, \ E \rightarrow \emptyset, \ \emptyset \rightarrow E\}$.

Note that the empty set appearing in (2.2) is a valid complex. It is used to model the inflow or outflow (or degradation) of molecules.

Having a notion of a reaction network in hand, we turn to the question of how to model the dynamical behavior of the counts of the different species. We describe the basic Markov model here and will consider deterministic dynamics at the end of the chapter. The precise connection between the two models will be considered in Chapter 4 when scaling limits of the stochastic models are considered.

Returning to Example 2.1 for the time being, let $X_1(t)$ and $X_2(t)$ be random variables giving the numbers of molecules of type A and B present in the system at time t, respectively. Denote by $R_1(t)$ and $R_2(t)$ the counting processes determining the number of times reactions (R1) and (R2) have occurred by time t. Clearly, X satisfies

$$X(t) = X(0) + R_1(t) \begin{pmatrix} -1 \\ 1 \end{pmatrix} + R_2(t) \begin{pmatrix} 1 \\ -1 \end{pmatrix}.$$

From the results of Chapter 1, the counting processes R_1 and R_2 can be specified by specifying their respective intensity functions. For the time being, we delay the conversation regarding what the appropriate form for these intensity functions should be and simply denote them by λ_1 and λ_2. The process X then satisfies the stochastic equation

$$X(t) = X(0) + Y_1\left(\int_0^t \lambda_1(X(s))ds\right) \begin{pmatrix} -1 \\ 1 \end{pmatrix} + Y_2\left(\int_0^t \lambda_2(X(s))ds\right) \begin{pmatrix} 1 \\ -1 \end{pmatrix},$$

where Y_1, Y_2 are independent, unit Poisson processes. We note that the generator (1.9) for the process of Example 2.1 is

$$Af(x) = \lambda_1(x)(f(x+\zeta_1) - f(x)) + \lambda_2(x)(f(x+\zeta_2) - f(x)), \quad x \in \mathbb{Z}_{\geq 0}^2,$$

where $\zeta_1 = -e_1 + e_2$, $\zeta_2 = e_1 - e_2$, and the forward or master equation (1.11) is

$$\dot{p}_x(t) = p_{x-\zeta_1}(t)\lambda_1(x-\zeta_1) + p_{x-\zeta_2}(t)\lambda_2(x-\zeta_2) - p_x(t)(\lambda_1(x) + \lambda_2(x)),$$

for $x \in \mathbb{Z}_{\geq 0}^2$.

Returning to the general reaction network of Definition 2.2, for each reaction $y_k \to y_k' \in \mathscr{R}$ we specify an intensity function $\lambda_k : \mathbb{Z}_{\geq 0}^n \to \mathbb{R}_{\geq 0}$. The number of times that the kth reaction occurs by time t can then be represented by the counting process

$$R_k(t) = Y_k\left(\int_0^t \lambda_k(X(s))ds\right),$$

where the Y_k are independent unit Poisson processes. The state of the system then satisfies the equation $X(t) = X(0) + \sum_k R_k(t)\zeta_k$, or

$$X(t) = X(0) + \sum_k Y_k\left(\int_0^t \lambda_k(X(s))ds\right)\zeta_k, \tag{2.3}$$

where the sum is over the reaction channels. (Recall that $\zeta_k = y_k' - y_k$.) The generator for the general model is

$$Af(x) = \sum_k \lambda_k(x)(f(x+\zeta_k) - f(x)), \tag{2.4}$$

where $f : \mathbb{Z}_{\geq 0}^d \to \mathbb{R}$ is any bounded function, whereas the Kolmogorov forward equation is

$$\dot{p}_x(t) = \sum_k \lambda_k(x-\zeta_k)p_{x-\zeta_k}(t) - \sum_k \lambda_k(x)p_x(t). \tag{2.5}$$

We now specify the intensity functions, or *kinetics*. The minimal assumption that can be put on the kinetics is that it is *stoichiometrically admissible*, which simply says that $\lambda_k(x) = 0$ if $x_i < y_{ki}$ for any $i \in \{1,\dots,n\}$. Stoichiometric admissibility ensures that reactions can only occur if there are sufficient molecules to produce the source complex and guarantees that the process remains within $\mathbb{Z}_{\geq 0}^n$ for all time. The most common choice of intensity function λ_k is that of stochastic *mass-action kinetics*. The stochastic form of the law of mass action says that for some constant κ_k, termed the *reaction rate constant*, the rate of the kth reaction should be

$$\lambda_k(x) = \kappa_k \prod_{i=1}^{n} y_{ki}! \binom{x}{y_k} = \kappa_k \prod_{i=1}^{n} \frac{x_i!}{(x_i - y_{ki})!}. \tag{2.6}$$

Note that the rate is proportional to the number of distinct subsets of the molecules present that can form the inputs for the reaction. This assumption reflects the idea that the system is well-stirred. The reaction rate constants are typically placed next to the arrow in the reaction diagram. The following table gives a representative list of reactions with their respective intensities under the assumption of mass-action kinetics,

Reaction	Intensity Function
$\emptyset \xrightarrow{\kappa_1} S_1$	$\lambda_1(x) = \kappa_1$
$S_1 \xrightarrow{\kappa_2} S_2$	$\lambda_2(x) = \kappa_2 x_1$
$S_1 + S_2 \xrightarrow{\kappa_3} S_3$	$\lambda_3(x) = \kappa_3 x_1 x_2$
$2S_1 \xrightarrow{\kappa_4} S_2$	$\lambda_4(x) = \kappa_4 x_1 (x_1 - 1)$

where similar expressions hold for intensity functions of higher order reactions.

2.1.1 Example: Gene transcription and translation

We give a series of stochastic models for gene *transcription* and *translation*. Transcription is the process by which the information encoded in a section of DNA is transferred to a piece of messenger RNA (mRNA). Next, this mRNA is translated by a ribosome, yielding proteins. We will give a series of three examples, with the first, Example 2.3, only including basic transcription, translation, and degradation of both mRNA and proteins. Next, in Example 2.4, we allow for the developed proteins to *dimerize*. Finally, in Example 2.5, we allow the resulting dimer to inhibit the production of the mRNA, and hence the protein and dimers themselves. This inhibition is an example of a *negative feedback loop* in that the protein product (i.e., the dimer) inhibits the rate of its own production. We note that each of our models leaves out many components, such as the RNA polymerase that is necessary for transcription, and the ribosomes that are critical for translation. Instead, we will assume that the abundances of such species are fixed and have been incorporated into the rate constants. More complicated, and realistic, models can of course incorporate both ribosomes and RNA polymerase.

Example 2.3. Consider a model of transcription and translation consisting of three species: $\mathscr{S} = \{G,M,P\}$, representing Gene, mRNA, and Protein, respectively. We suppose there are four possible transitions in our model:

$$
\begin{aligned}
R1) \quad & G \xrightarrow{\kappa_1} G+M && \text{(Transcription)}\\
R2) \quad & M \xrightarrow{\kappa_2} M+P && \text{(Translation)}\\
R3) \quad & M \xrightarrow{d_M} \varnothing && \text{(Degradation of mRNA)}\\
R4) \quad & P \xrightarrow{d_P} \varnothing && \text{(Degradation of protein)}
\end{aligned}
$$

where, as usual, the reaction rate constants for the different reactions have been placed above the reaction arrows.

We denote by $X(t) = (X_1(t), X_2(t), X_3(t)) \in \mathbb{Z}^3_{\geq 0}$ the vector giving the numbers of genes, mRNA molecules, and proteins at time t, respectively. The four reaction channels have reaction vectors

$$
\zeta_1 = \begin{bmatrix} 0 \\ 1 \\ 0 \end{bmatrix}, \quad
\zeta_2 = \begin{bmatrix} 0 \\ 0 \\ 1 \end{bmatrix}, \quad
\zeta_3 = \begin{bmatrix} 0 \\ -1 \\ 0 \end{bmatrix}, \quad
\zeta_4 = \begin{bmatrix} 0 \\ 0 \\ -1 \end{bmatrix},
$$

and respective intensities κ_1, $\kappa_2 X_1(t)$, $d_M X_1(t)$, $d_P X_2(t)$. The stochastic equation governing $X(t)$ is

$$
\begin{aligned}
X(t) = X(0) + Y_1(\kappa_1 t)\zeta_1 + Y_2\Big(\kappa_2 \int_0^t X_2(s)ds\Big)\zeta_2 + Y_3\Big(d_M \int_0^t X_2(s)ds\Big)\zeta_3 \\
+ Y_4\Big(d_P \int_0^t X_3(s)ds\Big)\zeta_4,
\end{aligned}
\tag{2.7}
$$

where Y_i, $i \in \{1,2,3,4\}$ are independent unit Poisson processes, and we have assumed that $X_1(t) \equiv 1$. Note that the rate of reaction 3 is zero when $X_2(t) = 0$ and the rate of reaction 4 is zero when $X_3(t) = 0$. Therefore, nonnegativity of the numbers of molecules is assured. See Figure 2.1 for a single realization of the stochastic model together with the associated deterministic model (see (2.9) below for a definition of the deterministic model of a chemical reaction system).   △

Example 2.4. We continue the previous example but now allow for the possibility that the protein dimerizes via the reaction $2P \xrightarrow{\kappa_3} D$. The degradation of the dimer is allowed by the reaction $D \xrightarrow{d_d} \varnothing$. The set of species for the model is now $\mathscr{S} = \{G,M,P,D\}$ and, keeping all other notation the same as in Example 2.3, we let $X_4(t)$ denote the number of dimers at time t. The stochastic equation for this model is

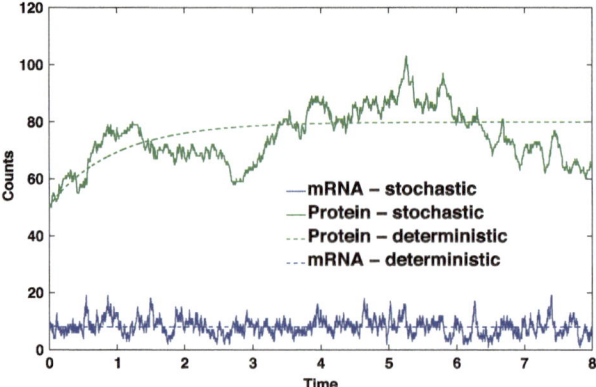

Fig. 2.1 A single trajectory of the model in Example 2.3 with choice of rate constants $\kappa_1 = 200$, $\kappa_2 = 10$, $d_M = 25$, and $d_p = 1$. The associated deterministic model with the same choice of rate constants is overlain (dashed).

$$X(t) = X(0) + Y_1(\kappa_1 t)\begin{bmatrix} 0 \\ 1 \\ 0 \\ 0 \end{bmatrix} + Y_2\left(\kappa_2 \int_0^t X_2(s)ds\right)\begin{bmatrix} 0 \\ 0 \\ 1 \\ 0 \end{bmatrix} + Y_3\left(d_M \int_0^t X_2(s)ds\right)\begin{bmatrix} 0 \\ -1 \\ 0 \\ 0 \end{bmatrix}$$

$$+ Y_4\left(d_P \int_0^t X_3(s)ds\right)\begin{bmatrix} 0 \\ 0 \\ -1 \\ 0 \end{bmatrix} + Y_5\left(\kappa_3 \int_0^t X_3(s)(X_3(s)-1)ds\right)\begin{bmatrix} 0 \\ 0 \\ -2 \\ 1 \end{bmatrix}$$

$$+ Y_6\left(d_d \int_0^t X_4(s)ds\right)\begin{bmatrix} 0 \\ 0 \\ 0 \\ -1 \end{bmatrix},$$

where $Y_i, i \in \{1,\dots,6\}$, are independent unit Poisson processes. See Figure 2.2 for a single realization of the stochastic process together with the associated deterministic model.   △

Example 2.5. Continuing the previous examples, we now allow for the dimer to interfere with, or *inhibit*, the production of the mRNA. Specifically, we assume the dimer can bind to the segment of DNA being translated, at which point no mRNA can be produced. Because the resulting dimers inhibit their own production (through the mRNA), this is an example of *negative feedback*. We must now explicitly model the gene to be in one of two states: bound and unbound. We let G remain the notation for the unbound gene, and use B to represent a bound gene. Let $X_5(t)$ give the number of bound genes at time t. Note that $X_1 + X_5$ give the total number of genes. We continue to assume that $X_1(t) + X_5(t) \equiv 1$. Now the set of species is

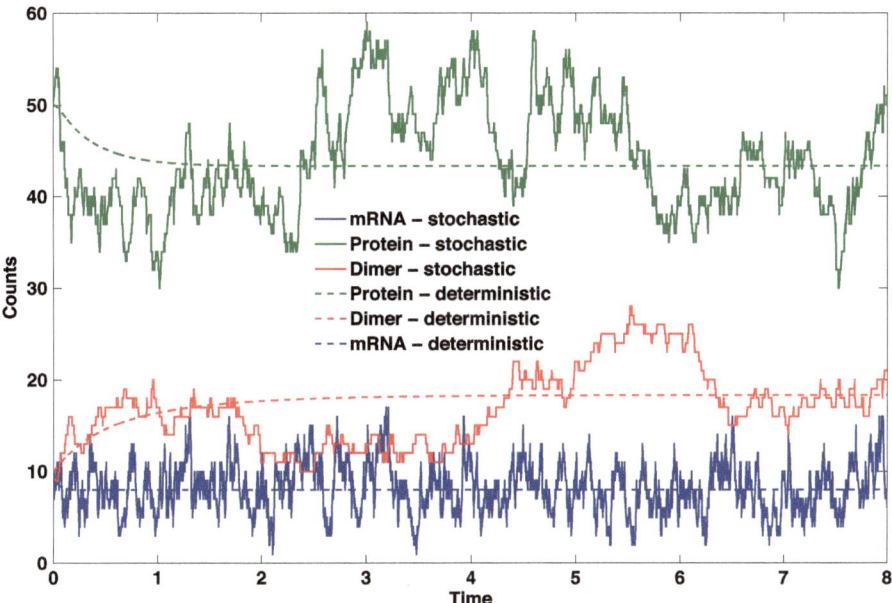

Fig. 2.2 A single trajectory of the model in Example 2.4 with choice of rate constants $\kappa_1 = 200$, $\kappa_2 = 10$, $d_M = 25$, $d_p = 1$, $\kappa_3 = 0.01$, and $d_d = 1$. The associated deterministic model with the same choice of rate constants is overlain (dashed).

$\mathscr{S} = \{G, M, P, D, B\}$ and we must add the reactions corresponding to binding and unbinding to our model,

$$G + D \underset{\kappa_{\text{off}}}{\overset{\kappa_{\text{on}}}{\rightleftharpoons}} B,$$

where $\kappa_{\text{on}}, \kappa_{\text{off}} > 0$ are the reaction rate constants. The stochastic equations are now

$$X(t) = X(0) + Y_1\left(\kappa_1 \int_0^t X_1(s)\,ds\right) \begin{bmatrix} 0 \\ 1 \\ 0 \\ 0 \\ 0 \end{bmatrix} + Y_2\left(\kappa_2 \int_0^t X_2(s)\,ds\right) \begin{bmatrix} 0 \\ 0 \\ 1 \\ 0 \\ 0 \end{bmatrix}$$

$$+ Y_3\left(d_M \int_0^t X_2(s)\,ds\right) \begin{bmatrix} 0 \\ -1 \\ 0 \\ 0 \\ 0 \end{bmatrix} + Y_4\left(d_P \int_0^t X_3(s)\,ds\right) \begin{bmatrix} 0 \\ 0 \\ -1 \\ 0 \\ 0 \end{bmatrix}$$

$$+Y_5\left(\kappa_5 \int_0^t X_3(s)(X_3(s)-1)ds\right)\begin{bmatrix}0\\0\\-2\\1\\0\end{bmatrix} + Y_6\left(d_d \int_0^t X_4(s)ds\right)\begin{bmatrix}0\\0\\0\\-1\\0\end{bmatrix}$$

$$+Y_7\left(\kappa_{on} \int_0^t X_4(s)X_1(s)ds\right)\begin{bmatrix}-1\\0\\0\\-1\\1\end{bmatrix} + Y_8\left(\kappa_{off} \int_0^t X_5(s)ds\right)\begin{bmatrix}1\\0\\0\\1\\-1\end{bmatrix}.$$

Note that the rate of the first reaction has changed to incorporate the fact that mRNA molecules will only be produced when the gene is free. We note that this example can be easily modified to have the dimer only slow the rate of production, or even raise the rate of production. If the rate of production is raised, then this would be an example with *positive feedback*. See Figure 2.3 for a single realization of the stochastic process modeled above (i.e., with the negative feedback) together with the associated deterministic model. Note the strikingly different behavior between the stochastic and deterministic model. Sample MATLAB code which produces realizations of this process is found online as a supplementary material. $\triangle$

2.1.2 Example: Virus kinetics

The following model of viral kinetics was first developed in [45] by Srivastava et al., and subsequently studied by Haseltine and Rawlings in [30], and Ball et al., in [10].

Example 2.6 (Viral infection). The model includes four time-varying species: the viral genome (G), the viral structural protein (S), the viral template (T), and the secreted virus itself (V). We denote these as species 1, 2, 3, and 4, respectively, and let $X_i(t)$ denote the number of molecules of species i at time t. The model has six reactions,

$$R_1: \quad T \xrightarrow{1} T+G, \qquad R_2: \quad G \xrightarrow{0.025} T, \qquad R_3: \quad T \xrightarrow{1000} T+S,$$

$$R_4: \quad T \xrightarrow{0.25} \emptyset, \qquad R_5: \quad S \xrightarrow{2} \emptyset, \qquad R_6: \quad G+S \xrightarrow{7.5\times 10^{-6}} V,$$

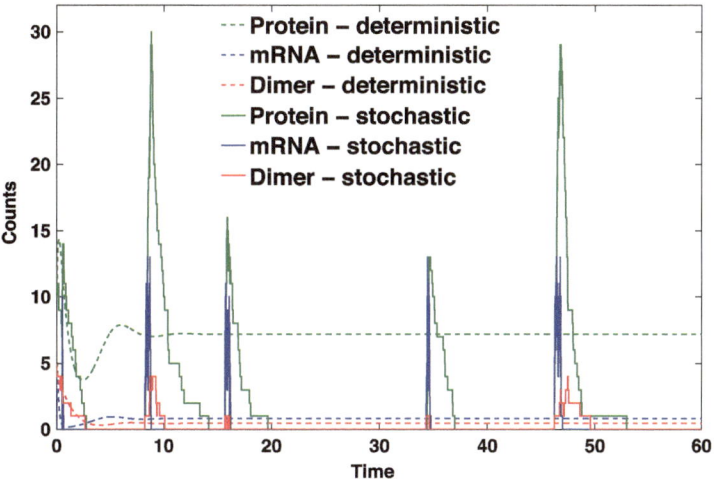

Fig. 2.3 A single trajectory of the model in Example 2.5 with choice of rate constants $\kappa_1 = 200$, $\kappa_2 = 10$, $d_M = 25$, $d_p = 1$, $\kappa_3 = 0.01$, $d_d = 1$, $k_{\text{on}} = 2$, $k_{\text{off}} = 0.1$. The associated deterministic model with the same choice of rate constants is overlain (dashed). Note that the negative feedback loop has allowed for strikingly different dynamics between the two models. Sample MATLAB code which produces realizations of this process is found online as a supplementary material.

where the units of time are in days. The stochastic equations for this model are

$$X_1(t) = X_1(0) + Y_1\left(\int_0^t X_3(s)ds\right) - Y_2\left(0.025\int_0^t X_1(s)ds\right)$$

$$- Y_6\left(7.5 \times 10^{-6}\int_0^t X_1(s)X_2(s)ds\right)$$

$$X_2(t) = X_2(0) + Y_3\left(1000\int_0^t X_3(s)ds\right) - Y_5\left(2\int_0^t X_2(s)ds\right)$$

$$- Y_6\left(7.5 \times 10^{-6}\int_0^t X_1(s)X_2(s)ds\right)$$

$$X_3(t) = X_3(0) + Y_2\left(0.025\int_0^t X_1(s)ds\right) - Y_4\left(0.25\int_0^t X_3(s)ds\right)$$

$$X_4(t) = X_4(0) + Y_6\left(7.5 \times 10^{-6}\int_0^t X_1(s)X_2(s)ds\right). \tag{2.8}$$

Note that the rate constants of the above model vary over several orders of magnitude, which will in turn cause a large variation in the molecular counts of the different species. See Figure 2.4 for a single realization of this process.   △

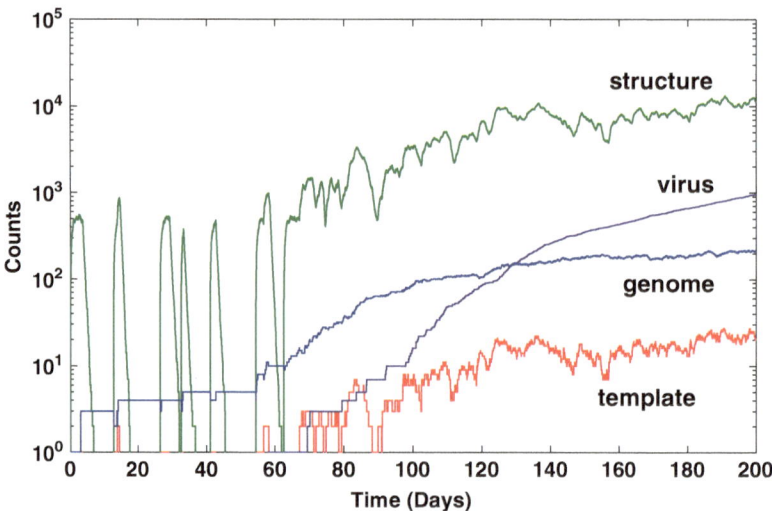

Fig. 2.4 A single trajectory of the model in Example 2.6. Note that the y-axis uses a log scale.

2.1.3 Example: Enzyme kinetics

We consider a standard model in which an enzyme catalyzes the conversion of some substrate to a product.

Example 2.7. Let S be a substrate, E be an enzyme, SE be an enzyme-substrate complex, and P be a product and consider the reaction network

$$S + E \underset{\kappa_2}{\overset{\kappa_1}{\rightleftharpoons}} SE \overset{\kappa_3}{\to} P + E,$$

which is a slightly simplified version of (2.2). Letting X_1, X_2, X_3, X_4 be the processes giving the counts of species S, E, SE, and P, respectively, the stochastic equations for this model are

$$X_1(t) = X_1(0) - Y_1\Big(\int_0^t \kappa_1 X_1(s) X_2(s) ds\Big) + Y_2\Big(\int_0^t \kappa_2 X_3(s) ds\Big)$$

$$X_2(t) = X_2(0) - Y_1\Big(\int_0^t \kappa_1 X_1(s) X_2(s) ds\Big) + Y_2\Big(\int_0^t \kappa_2 X_3(s) ds\Big) + Y_3\Big(\int_0^t \kappa_3 X_3(s) ds\Big)$$

$$X_3(t) = X_3(0) + Y_1\Big(\int_0^t \kappa_1 X_1(s) X_2(s) ds\Big) - Y_2\Big(\int_0^t \kappa_2 X_3(s) ds\Big) - Y_3\Big(\int_0^t \kappa_3 X_3(s) ds\Big)$$

$$X_4(t) = X_4(0) + Y_3\Big(\int_0^t \kappa_3 X_3(s) ds\Big).$$

We will study stationary distributions for variants of this model in Chapter 3, and derive Michaelis-Menten kinetics from a multi-scale analysis of this model in Chapter 4. $\triangle$

2.2 Deterministic models of biochemical reaction systems, and first-order reaction networks

Consider a chemical reaction network $\{\mathscr{S},\mathscr{C},\mathscr{R}\}$ in which the counts of the constituent species is so high that a deterministic model accurately captures the dynamics of their concentrations (see Chapter 4). Let $x(t) \in \mathbb{R}^n_{>0}$ be the vector for which $x_i(t)$ models the concentration of species S_i at time t. Then, recalling that for the kth reaction, $y_k \to y'_k$, we take $\zeta_k = y'_k - y_k \in \mathbb{Z}^d$, the most common choice for the deterministically modeled system is

$$\dot{x}(t) = \sum_k \kappa_k x(t)^{y_k} \zeta_k, \tag{2.9}$$

where for two vectors $u, v \in \mathbb{R}^d_{\geq 0}$, we define $u^v = \prod_{i=1}^d u_i^{v_i}$ with $0^0 \equiv 1$. The choice of rate function $\kappa_k x^{y_k}$ is called *deterministic mass-action kinetics*.

We delay until Chapter 4 the arguments for how the deterministic model can be realized as a scaling limit of the stochastic model. Here, we simply consider a special situation in which the ODE model (2.9) accurately captures the mean behavior of the corresponding stochastic model.

Definition 2.8. We say a reaction network $\{\mathscr{S},\mathscr{C},\mathscr{R}\}$ is a *first-order reaction network* if for each $y_k \to y'_k \in \mathscr{R}$, we have $|y_k| \in \{0,1\}$.

That is, a network is a first-order reaction network if reactions are only of the form

$$\emptyset \to *, \quad \text{or} \quad S_i \to *,$$

where $*$ can represent any linear combination of the species. Note that Example 2.3 is a first-order reaction network, whereas Examples 2.1, 2.4, 2.5, and 2.6 are not.

It is relatively easy to show that if $\{\mathscr{S},\mathscr{C},\mathscr{R}\}$ is a first-order reaction network then $E[X(t)]$ is a solution to the system (2.9), where κ_k are the reaction rate constants for the stochastic model. The key observation is that in the case that the system is first order, the intensity functions are linear. For example, in the case that the kth reaction is of the form $\emptyset \to *$, the associated intensity function is simply κ_k, whereas in the case that the kth reaction is of the form $S_{i_k} \to *$, the associated intensity function is $\kappa_k X_{i_k}$. In either case we may conclude that $E[\lambda_k(X(s))] = \lambda_k(E[X(s)])$. Letting R_k be the counting process for the kth reaction channel, we know from Chapter 1 that $R_k(t) - \int_0^t \lambda_k(X(s))ds$ is a martingale and so,

$$E[X_i(t)] = E[X_i(0)] + E[\sum_k R_k(t)\zeta_{ki}]$$

$$= E[X_i(0)] + E[\int_0^t \sum_k \lambda_k(X(s))\zeta_{ki}\,ds] \tag{2.10}$$

$$= E[X_i(0)] + \int_0^t \sum_k \lambda_k(E[X(s)])\zeta_{ki}\,ds,$$

and differentiating shows the assertion to be true. Second, and higher, moments can be calculated in a similar manner.

Example 2.9. Consider the reaction network

$$\emptyset \underset{\mu}{\overset{\lambda}{\rightleftharpoons}} S,$$

where $\lambda, \mu > 0$. The reaction $\emptyset \to S$ models the arrival of molecules S, whereas the reaction $S \to \emptyset$ models their degradation. Note that the corresponding Markov model is the same as that for an $M/M/\infty$ queue.

Let $X(t)$ denote the number of molecules of S at time $t \geq 0$, and let A be the generator for the stochastically modeled system. Assuming $X(0) = x$ is deterministic, and taking $f(y) = y$ to be the identity function, we have

$$E[X(t)] = E[f(X(t))] = x + E[\int_0^t Af(X(s))ds]$$

$$= x + E[\int_0^t \left(\lambda(X(s) + 1 - X(s))\right) + \mu X(s)(X(s) - 1 - X(s))ds]$$

$$= x + \int_0^t (\lambda - \mu E[X(s)])ds.$$

Solving yields

$$E[X(t)] = xe^{-\mu t} + \frac{\lambda}{\mu}(1 - e^{-\mu t}). \tag{2.11}$$

The second moment, and hence the variance, of the process can be calculated in a similar manner. Letting $f(y) = y^2$, we have

$$E[X(t)^2] = E[f(X(t))] = E[\int_0^t Af(X(s))ds]$$

$$= x^2 + E[\int_0^t \lambda((X(s) + 1)^2 - X(s)^2) + \mu X(s)((X(s) - 1)^2 - X(s)^2)\,ds]$$

$$= x^2 + E[\int_0^t \lambda(2X(s) + 1) + \mu X(s)(-2X(s) + 1)\,ds]$$

$$= x^2 + (2\lambda + \mu)\int_0^t E[X(s)]ds + \lambda t - 2\mu \int_0^t E[X(s)^2]\,ds.$$

Using (2.11), we conclude that $E[X(t)^2]$ satisfies

$$E[X(t)^2] = x^2 + \frac{(2\lambda + \mu)(-e^{-\mu t}\mu x + \lambda t\mu + e^{-\mu t}\lambda + \mu x - \lambda)}{\mu^2} + \lambda t$$

$$- 2\mu \int_0^t E[X(s)^2]\,ds.$$

Solving the above equation yields

$$E[X(t)^2] = \left(1 + 2\frac{\lambda}{\mu}\right)xe^{-\mu t} + \left(1 - \frac{\lambda}{\mu}\right)e^{-\mu t} + \left(1 - 2e^{-\mu t}\right)\frac{\lambda^2}{\mu^2}$$
$$+ e^{-2\mu t}\left(x^2 - 2\frac{\lambda x}{\mu} - x + \frac{\lambda^2}{\mu^2}\right).$$

Solving for the variance we find

$$\mathrm{Var}(X(t)) = E[X(t)^2] - E[X(t)]^2$$
$$= \frac{\lambda}{\mu}\left(1 - e^{-\mu t}\right) + xe^{-\mu t}(1 - e^{-\mu t}).$$

Note that for this example $\lim_{t\to\infty}\mathrm{Var}(X(t)) = \frac{\lambda}{\mu} = \lim_{t\to\infty}EX(t)$, a fact that will be explained in more detail in Chapter 3. $\triangle$

Problems

2.1. Consider the reaction network

$$\emptyset \xrightarrow{\kappa_1} S_1 \xrightarrow{\kappa_2} S_2 \xrightarrow{\kappa_3} \emptyset.$$

Compute the expectations $E[X_1(t)]$, $E[X_2(t)]$.

2.2. Consider the reaction network for a pure birth process

$$S \xrightarrow{\lambda} 2S,$$

where $\lambda > 0$. Compute $E[X(t)]$ and $\mathrm{Var}(X(t))$.

2.3. Consider the reaction network for a birth and death process

$$\emptyset \xleftarrow{\mu} S \xrightarrow{\lambda} 2S,$$

where $\lambda, \mu > 0$. Compute $E[X(t)]$ and $\mathrm{Var}(X(t))$.

2.4. Consider the reaction network introduced in Example 2.9:

$$\emptyset \underset{\mu}{\overset{\lambda}{\rightleftharpoons}} S,$$

where $\lambda, \mu > 0$. Derive the equations for the time evolution of the mean of the process by noting that $E[X(t)] = \sum_{k=0}^{\infty} kP\{X(t) = k\}$, differentiating, and using the Kolmogorov forward equation (2.5).

Chapter 3
Stationary distributions of stochastically modeled reaction systems

3.1 Introduction

One of the main objects of study in the analysis of deterministically modeled chemical reaction systems are the fixed points for the system of equations (2.9). In the deterministic modeling context, key questions related to fixed points include:

- For what class of models are there fixed points, and how many are there?
- When there are fixed points, what can we say about their local or global stability?

The second question has received a remarkable amount of attention over the past ten years as it relates to one of the main open problems in the field, the so-called *Global Attractor Conjecture*, which states that the equilibria for a deterministically modeled system satisfying certain easily checked criteria (as detailed in Theorem 3.5) are all globally asymptotically stable relative to their invariant manifolds [3, 16]. The right-hand side of (2.9) is a polynomial, and so it is not surprising that algebraic methods have been successfully employed in a number of areas related to fixed points of mass-action systems [16, 39, 41].

Analogous questions are asked about the stationary behavior of the corresponding stochastic model:

- When will a chemical system admit a stationary distribution that charges all states of the state space?
- Can we characterize the resulting stationary distribution?

Of course, simple example networks such as $A \to \emptyset$, whose corresponding stochastic model converges to zero, or $\emptyset \to A$, whose corresponding stochastic model increases without bound, show that not all reaction networks admit stationary distributions that charge all states. In particular, see [8] for a large class of networks with stationary distributions that only charge states on the boundary of $\mathbb{Z}^n_{\geq 0}$.

© Springer International Publishing Switzerland 2015
D.F. Anderson, T.G. Kurtz, *Stochastic Analysis of Biochemical Systems*,
Mathematical Biosciences Institute Lecture Series 1,
DOI 10.1007/978-3-319-16895-1_3

3.2 Network conditions, complex-balanced equilibria, and the deficiency zero theorem

It is clear that the network $\emptyset \to A$ induces both a deterministic and stochastic model that grows unboundedly in time. Just as obvious is the fact that the network $A \to \emptyset$ induces models that decay to zero as time goes to infinity. Significantly less obvious is the long-time behavior of the dynamical models associated with more complicated networks such as

$$S + E \rightleftharpoons SE \rightleftharpoons P + E, \quad E \rightleftharpoons \emptyset, \tag{3.1}$$

which is a variant of the model of Example 2.7. The following definitions and results help us relate properties of the network architecture to those of the associated dynamical systems. Most of the terminology and results in this section stem from the works of Horn, Jackson, and Feinberg [21, 32]. See also [20, 29] for survey articles.

Definition 3.1. A chemical reaction network, $\{\mathscr{S}, \mathscr{C}, \mathscr{R}\}$, is called *weakly reversible* if for any reaction $y_k \to y_k' \in \mathscr{R}$, there is a sequence of directed reactions beginning with y_k' as a source complex and ending with y_k as a product complex. That is, there exist complexes $y_1, \ldots, y_r$ such that $y_k' \to y_1, y_1 \to y_2, \ldots, y_r \to y_k \in \mathscr{R}$. A network is called *reversible* if $y_k' \to y_k \in \mathscr{R}$ whenever $y_k \to y_k' \in \mathscr{R}$.

Note, for example, that the reaction network (3.1) is reversible, and hence weakly reversible, whereas the network

$$S + E \rightleftharpoons SE \to P + E, \quad E \rightleftharpoons \emptyset,$$

is neither reversible nor weakly reversible. The network

$$
\begin{array}{ccc}
A & \longrightarrow & B \\
& \nwarrow \quad \swarrow & \\
& C &
\end{array}
\tag{3.2}
$$

is weakly reversible, but not reversible.

Now consider the unique directed graph implied by a reaction network. That is, consider the graphs of the form (3.1) and (3.2). We term each connected component of the resulting graph a *linkage class* and denote the number of linkage classes by $\ell \in \mathbb{Z}_{\geq 0}$. For example, the reaction network (3.1) has two linkage classes, whereas the reaction network (3.2) has only one.

It is easy to see that a chemical reaction network is weakly reversible if and only if each of the linkage classes of its graph is strongly connected.

Definition 3.2. $S = \text{span}_{\{y_k \to y_k' \in \mathscr{R}\}} \{y_k' - y_k\}$ is the *stoichiometric subspace* of the network. For $c \in \mathbb{R}^n$ we say $c + S$ and $(c + S) \cap \mathbb{R}_{\geq 0}^n$ are the *stoichiometric compatibility classes* and *nonnegative stoichiometric compatibility classes* of the network, respectively. Denote $\dim(S) = s$.

For example, the network

$$A \rightleftharpoons B, \quad 2A \rightleftharpoons 2B \tag{3.3}$$

has reaction vectors

$$\begin{pmatrix} -1 \\ 1 \end{pmatrix}, \begin{pmatrix} 1 \\ -1 \end{pmatrix}, \begin{pmatrix} -2 \\ 2 \end{pmatrix}, \begin{pmatrix} 2 \\ -2 \end{pmatrix}$$

and stoichiometric subspace $\{x \in \mathbb{R}^2 : x_1 + x_2 = 0\}$, whereas the network

$$\emptyset \rightleftharpoons A \rightleftharpoons B, \quad 2A \rightleftharpoons 2B$$

has reaction vectors

$$\begin{pmatrix} 1 \\ 0 \end{pmatrix}, \begin{pmatrix} -1 \\ 0 \end{pmatrix}, \begin{pmatrix} -1 \\ 1 \end{pmatrix}, \begin{pmatrix} 1 \\ -1 \end{pmatrix}, \begin{pmatrix} -2 \\ 2 \end{pmatrix}, \begin{pmatrix} 2 \\ -2 \end{pmatrix}$$

and stoichiometric subspace $\mathbb{R}^2$. It is straightforward to show that for both stochastic and deterministic models, the state of the system remains within a single stoichiometric compatibility class for all time. For example, letting R_k be the counting processes giving the number of times the kth reaction has taken place, we see

$$X(t) - X(0) = \sum_k R_k(t)\zeta_k \in S \implies X(t) \in X(0) + S,$$

for all $t > 0$. Note that for a given nonnegative initial condition $X(0) \in \mathbb{Z}_{\geq 0}^n$, the state space for the stochastic model will be a subset of the set $(X(0) + S) \cap \mathbb{Z}_{\geq 0}^n$ so long as the kinetics are stoichiometrically admissible.

Definition 3.3. The *deficiency* of a chemical reaction network, $\{\mathscr{S}, \mathscr{C}, \mathscr{R}\}$, is $\delta = |\mathscr{C}| - \ell - s$, where $|\mathscr{C}|$ is the number of complexes, ℓ is the number of linkage classes of the network graph, and s is the dimension of the stoichiometric subspace of the network.

For example, the network (3.3) consists of four complexes, two linkage classes, and the dimension of the stoichiometric compatibility class is one. Hence, the deficiency is $4 - 2 - 1 = 1$. On the other hand, the network (3.2) consists of three complexes, one linkage class, and the dimension of the stoichiometric compatibility class is two. Thus, the deficiency is $3 - 1 - 2 = 0$. Zero is a lower bound on the deficiency of a network.

Proposition 3.4. *The deficiency of a network is nonnegative.*

Proof. Let $C_i \subset \mathscr{C}$ be the complexes in the ith linkage class, and let S_i be the span of the reaction vectors of the edges in the ith linkage class. Then $\dim(S_i) \leq |C_i| - 1$ and,

$$\dim(S) \leq \sum_{i=1}^{\ell} \dim(S_i) \leq \sum_{i=1}^{\ell} |C_i| - \ell = |\mathscr{C}| - \ell.$$

$\square$

An equilibrium, $c \in \mathbb{R}_{\geq 0}^n$, of the deterministic model (2.9) is said to be a *complex-balanced* equilibrium if for each complex $\eta \in \mathscr{C}$,

$$\sum_{\{k:\eta=y_k\}} \kappa_k c^{y_k} = \sum_{\{k:\eta=y_k'\}} \kappa_k c^{y_k}, \tag{3.4}$$

where the sum on the left is over reactions for which η is the source complex and the sum on the right is over those reactions for which η is the product complex.

See [20] or [29] for a proof of the following.

Theorem 3.5. *Let $\{\mathscr{S},\mathscr{C},\mathscr{R}\}$ be a chemical reaction network with deterministic mass-action kinetics. Suppose there is a complex-balanced equilibrium, $c \in \mathbb{R}_{>0}^n$, satisfying (3.4). Then, there is precisely one equilibrium in the interior of each non-negative stoichiometric compatibility class and each such equilibrium is complex-balanced.*

Hence, it makes sense to talk about a complex-balanced system. The following result by Feinberg makes a connection between the deficiency of a network and its ability to admit complex-balanced equilibria [20, 21].

Theorem 3.6. *Let $\{\mathscr{S},\mathscr{C},\mathscr{R}\}$ be a chemical reaction network with deterministic mass-action kinetics. If the network has a deficiency of zero, then there exists a complex-balanced equilibrium $c \in \mathbb{R}_{>0}^n$ if and only if the network is weakly reversible.*

3.3 Stationary distributions for complex-balanced models

Denote the closed, irreducible components of the state space of a countable Markov chain by $\{\Gamma\}$. All stationary distributions of the chain can be written as

$$\pi = \sum_{\Gamma} \alpha_\Gamma \pi_\Gamma, \tag{3.5}$$

where $\alpha_\Gamma \geq 0$, $\sum_\Gamma \alpha_\Gamma = 1$, and where π_Γ is the unique stationary distribution satisfying $\pi_\Gamma(\Gamma) = 1$, for those Γ for which a stationary distribution exists.

The following result and proof first appeared in [6], though earlier relevant work is found in [35].

Theorem 3.7. *Let $\{\mathscr{S},\mathscr{C},\mathscr{R}\}$ be a chemical reaction network and let $\{\kappa_k\}$ be a choice of rate constants. Suppose that, modeled deterministically via (2.9), the system is complex-balanced with complex-balanced equilibrium $c \in \mathbb{R}_{>0}^n$. Then the stochastically modeled system with intensities (2.6) has a stationary distribution consisting of the product of Poisson distributions,*

$$\pi(x) = \prod_{i=1}^n \frac{c_i^{x_i}}{x_i!} e^{-c_i}, \qquad x \in \mathbb{Z}_{\geq 0}^n. \tag{3.6}$$

If $\mathbb{Z}_{\geq 0}^n$ is irreducible, then (3.6) is the unique stationary distribution, whereas if $\mathbb{Z}_{\geq 0}^n$ is not irreducible, then the π_Γ of equation (3.5) are given by the product-form stationary distributions

$$\pi_\Gamma(x) = M_\Gamma \prod_{i=1}^n \frac{c_i^{x_i}}{x_i!}, \qquad x \in \Gamma,$$

and $\pi_\Gamma(x) = 0$ otherwise, where M_Γ is a positive normalizing constant.

Remark 3.8. It is important to note that the same rate constants $\{\kappa_k\}$ are used for both the deterministic and stochastic models.

Proof. Let π satisfy (3.6) where $c \in \mathbb{R}_{>0}^n$ is a complex-balanced equilibrium. The forward equation (1.11) implies π is a stationary distribution if it simultaneously satisfies

$$\sum_\ell \lambda_\ell(x - \zeta_\ell)\pi(x - \zeta_\ell) - \sum_\ell \lambda_\ell(x)\pi(x) = 0, \tag{3.7}$$

for each $x \in \mathbb{Z}_{\geq 0}^n$, and $\sum_{x \in \mathbb{Z}_{\geq 0}^n} \pi(x) = 1$. The second condition holds automatically, so we just need to check the first. Plugging π and (2.6) into equation (3.7) and simplifying yields

$$\sum_k \kappa_k c^{y_k - y_k'} \frac{1}{(x - y_k')!} \prod_{\ell=1}^n 1_{\{x_\ell \geq y_{\ell k}'\}} = \sum_k \kappa_k \frac{1}{(x - y_k)!} \prod_{\ell=1}^n 1_{\{x_\ell \geq y_{\ell k}\}}. \tag{3.8}$$

The key step in the proof is to note that we may write the above sums as

$$\sum_k \kappa_k c^{y_k - y_k'} \frac{1}{(x - y_k')!} \prod_{\ell=1}^n 1_{\{x_\ell \geq y_{\ell k}'\}} = \sum_{\eta \in \mathscr{C}} \sum_{\{k:y_k'=\eta\}} \kappa_k c^{y_k - y_k'} \frac{1}{(x - y_k')!} \prod_{\ell=1}^n 1_{\{x_\ell \geq y_{\ell k}'\}}$$

and

$$\sum_k \kappa_k \frac{1}{(x - y_k)!} \prod_{\ell=1}^n 1_{\{x_\ell \geq y_{\ell k}\}} = \sum_{\eta \in \mathscr{C}} \sum_{\{k:y_k=\eta\}} \kappa_k \frac{1}{(x - y_k)!} \prod_{\ell=1}^n 1_{\{x_\ell \geq y_{\ell k}\}},$$

where the enumeration $\{k : y_k' = \eta\}$ is over reactions for which η is the product complex and the enumeration $\{k : y_k = \eta\}$ is over reactions for which η is the source complex. Hence, equation (3.8) will be satisfied if for each complex $\eta \in \mathscr{C}$,

$$\sum_{\{k:y_k'=\eta\}} \kappa_k c^{y_k - \eta} \frac{1}{(x - \eta)!} \prod_{\ell=1}^n 1_{\{x_\ell \geq \eta_\ell\}} = \sum_{\{k:y_k=\eta\}} \kappa_k \frac{1}{(x - \eta)!} \prod_{\ell=1}^n 1_{\{x_\ell \geq \eta_\ell\}}. \tag{3.9}$$

Both the state x and the complex η are fixed in the above equation, and so multiplying (3.9) through by $c^\eta(x - \eta)! > 0$ yields the equivalent equation

$$\sum_{\{k:y_k'=\eta\}} \kappa_k c^{y_k} = \sum_{\{k:y_k=\eta\}} \kappa_k c^\eta = \sum_{\{k:y_k=\eta\}} \kappa_k c^{y_k},$$

which is simply (3.4).

To complete the proof, one need only observe that the normalized restriction of π to any closed, irreducible subset Γ must also be a stationary distribution. $\qquad\square$

Note that in light of Theorem 3.6, the conclusions of Theorem 3.7 hold *for any choice of rate constants* if the network is weakly reversible and has a deficiency of zero.

We present a series of examples demonstrating the usefulness of Theorem 3.7.

Example 3.9. Consider the network of Example 2.9,

$$\emptyset \underset{\mu}{\overset{\lambda}{\rightleftharpoons}} S,$$

where $\lambda, \mu > 0$. This network has a deficiency of $\delta = 2 - 1 - 1 = 0$, and is weakly reversible. The equilibrium of the corresponding deterministically modeled system is $c = \lambda/\mu$ and the state space is all of $\mathbb{Z}_{\geq 0}$. Hence, the stationary distribution of the stochastic model is Poisson with parameter λ/μ. Note that this model is identical to the simplest model of an infinite server queue. This example along with the next two is well-known in the queueing literature as is the form of their stationary distributions. $\qquad\triangle$

Example 3.10. Consider the network

$$S_1 \underset{\kappa_2}{\overset{\kappa_1}{\rightleftharpoons}} S_2,$$

where $\kappa_1, \kappa_2 > 0$. Suppose that $X_1(0) + X_2(0) = N$ so that $X_1(t) + X_2(t) = N$ for all $t > 0$. This system has a deficiency of zero and is weakly reversible. A complex-balanced equilibrium to the deterministically modeled system is

$$c = \left(\frac{\kappa_2}{\kappa_1 + \kappa_2}, \frac{\kappa_1}{\kappa_1 + \kappa_2} \right),$$

and the product-form stationary distribution for the system is therefore

$$\pi(x) = M \frac{c_1^{x_1} c_2^{x_2}}{x_1! \, x_2!},$$

where $M > 0$ is a normalizing constant. Using that $X_1(t) + X_2(t) = N$ for all t yields

$$\pi_1(x_1) = M \frac{c_1^{x_1}}{x_1!} \frac{c_2^{N-x_1}}{(N-x_1)!} = \frac{M}{x_1!(N-x_1)!} c_1^{x_1} (1-c_1)^{N-x_1},$$

for $0 \leq x_1 \leq N$. After setting $M = N!$, we see that X_1 is binomially distributed. Similarly,

$$\pi_2(x_2) = \binom{N}{x_2} c_2^{x_2} (1-c_2)^{N-x_2},$$

for $0 \leq x_2 \leq N$. Note that due to the conservation relation, X_1 and X_2 are *not* independent under this stationary distribution. $\qquad\triangle$

The following example shows how to generalize the conclusions of the previous two.

Example 3.11 (Stationary distributions for first-order reaction networks). We denote by *SSC* (for *single species complexes*) the class of first-order reaction networks for which all complexes consist of either a single species or the empty set. For example, the network

$$\emptyset \leftarrow S \rightarrow 2S,$$

is a first-order reaction network, but is *not* contained in *SSC*. It is known that a weakly reversible *SSC* network admits a product-form stationary distribution of the form (3.6) [22, 35]. We show here how this fact follows trivially from Theorem 3.7.

First, it is relatively straightforward to show that all *SSC* networks have a deficiency of zero (see Problem 3.2). Therefore, Theorem 3.7 is applicable to all *SSC* networks that are weakly reversible. Consider such a reaction network with only one linkage class (if there is more than one linkage class, we can consider the different linkage classes as distinct networks/systems).

We say that the network is *open* if $\emptyset \in \mathscr{C}$, and otherwise say it is *closed*. In the case of an open, weakly reversible, *SSC* network we see that $S = \mathbb{R}^n$, the state space is all of $\mathbb{Z}_{\geq 0}^n$, and $\mathbb{Z}_{\geq 0}^n$ is irreducible. Thus, by Theorem 3.7 the unique stationary distribution is

$$\pi(x) = \prod_{i=1}^{n} \frac{c_i^{x_i}}{x_i!} e^{-c_i}, \qquad x \in \mathbb{Z}_{\geq 0}^n,$$

where $c \in \mathbb{R}_{>0}^n$ is the complex-balanced equilibrium of the associated (linear) deterministic system. Therefore, when in distributional equilibrium, the species numbers are independent and have Poisson distributions.

Now suppose the network is closed, weakly reversible, and is a single linkage class *SSC* network. Suppose further that $X_1(0) + \cdots + X_n(0) = N$. Then, $\Gamma_N = \{x \in \mathbb{Z}_{\geq 0}^n : x_1 + \cdots + x_n = N\}$ is closed and irreducible (see Problem 3.3). By Theorem 3.7, in distributional equilibrium $X(t)$ has a multinomial distribution. That is, for any $x \in \mathbb{Z}_{\geq 0}^n$ satisfying $x_1 + x_2 + \cdots + x_n = N$

$$\pi_\Gamma(x) = \binom{N}{x_1, x_2, \ldots, x_n} c^x = \frac{N!}{x_1! \cdots x_n!} c_1^{x_1} \cdots c_n^{x_n}, \tag{3.10}$$

where $c \in \mathbb{R}_{>0}^n$ is the equilibrium of the associated deterministic system normalized so that $\sum_i c_i = 1$. $\triangle$

We now turn our attention to reaction systems with nonlinear intensity functions.

Example 3.12. We modify the model of Example 2.7 and consider the open network

$$S + E \rightleftharpoons ES \rightleftharpoons P + E, \qquad E \rightleftharpoons \emptyset \rightleftharpoons S. \tag{3.11}$$

The network (3.11) is reversible and has six complexes and two linkage classes. The dimension of the stoichiometric subspace is readily checked to be four, and so the network has a deficiency of zero. Theorem 3.7 applies and so the stochastically modeled system has a product-form stationary distribution of the form (3.6). It is readily checked that the unique closed, irreducible communicating class of the stochastically modeled system is all of $\mathbb{Z}_{\geq 0}^4$, and so in equilibrium the species are independent and have Poisson distributions. $\triangle$

Example 3.13. We again modify Example 2.7 and consider now the open network

$$S+E \underset{\kappa_{-1}}{\overset{\kappa_1}{\rightleftharpoons}} ES \underset{\kappa_{-2}}{\overset{\kappa_2}{\rightleftharpoons}} P+E, \qquad \emptyset \underset{\kappa_{-3}}{\overset{\kappa_3}{\rightleftharpoons}} E, \qquad (3.12)$$

where a choice of rate constants has now been made explicit. The network is reversible, there are five complexes, two linkage classes, and the dimension of the stoichiometric compatibility class is three. Therefore, Theorem 3.7 implies that the stochastically modeled system has a product-form stationary distribution of the form (3.6). Unlike in Example 3.12, there is now a conserved quantity

$$X_S(t) + X_{ES}(t) + X_P(t) = N,$$

where $N > 0$. Therefore, after solving for the normalizing constant, we have that for any $x \in \mathbb{Z}_{\geq 0}^4$ satisfying $x_2 + x_3 + x_4 = N$

$$\pi(x) = e^{-c_1} \frac{c_1^{x_1}}{x_1!} \frac{N!}{x_2! x_3! x_4!} c_2^{x_2} c_3^{x_3} c_4^{x_4} = e^{-c_1} \frac{c_1^{x_1}}{x_1!} \binom{N}{x_2, x_3, x_4} c_2^{x_2} c_3^{x_3} c_4^{x_4},$$

where $c = (\kappa_3/\kappa_{-3}, c_2, c_3, c_4)$ has been chosen so that $c_2 + c_3 + c_4 = 1$. Thus, when the stochastically modeled system is in distributional equilibrium we have that: (a) E has a Poisson distribution with parameter κ_3/κ_{-3}, (b) S, ES, and P are multinomially distributed, and (c) E is independent from S, ES, and P. $\triangle$

Problems

3.1. What are the deficiencies of the models in Examples 2.3 and 2.6?

3.2. Show that all *SSC* networks have a deficiency of zero. Give an example of a first order network with a deficiency of one.

3.3. Let $\{\mathscr{S}, \mathscr{C}, \mathscr{R}\}$ be a closed *SSC* network with only one linkage class. Show that $\Gamma_N = \{x \in \mathbb{Z}_{\geq 0}^n : x_1 + \cdots + x_n = N\}$ is a closed and irreducible component of the state space.

3.4. Find the stationary distribution for the stochastic model associated with the reaction network

$$2P \underset{\kappa_2}{\overset{\kappa_1}{\rightleftharpoons}} D,$$

and initial condition $X_P(0), X_D(0)$. Note that this is the standard model for protein *dimerization*.

3.5. Find the stationary distribution of the model with network

$$\emptyset \underset{\kappa_{-1}}{\overset{\kappa_1}{\rightleftharpoons}} S_1, \quad 2S_1 \underset{\kappa_{-2}}{\overset{\kappa_2}{\rightleftharpoons}} S_2.$$

3.6. Find the stationary distribution of the model with network

$$A \underset{\kappa_{-1}}{\overset{\kappa_1}{\rightleftharpoons}} 2A.$$

Chapter 4
Analytic approaches to model simplification and approximation

4.1 Limits under the classical scaling

The state vectors of the models of reaction systems introduced in Chapter 2 give the numbers of molecules of the chemical species in the system. In classical chemistry, the state of a reaction system is usually described in terms of chemical concentrations rather than numbers of molecules. These descriptions are, of course, related, the chemical concentration of a species being the number of molecules of the species divided by Avogadro's number N_A ($\approx 6 \times 10^{23}$) times the volume v of the reaction mixture.

For a well-mixed, binary reaction,

$$S_1 + S_2 \to *, \tag{4.1}$$

the reaction rate should vary inversely with the volume, so we set $N_v = N_A v$ and assume that the rate function for (4.1) has the form

$$\frac{\kappa}{N_v} x_1 x_2, \tag{4.2}$$

where x_1 and x_2 are the numbers of molecules of species S_1 and S_2, respectively. If we rewrite (4.2) in terms of the concentrations $c_i = N_v^{-1} x_i$, we have

$$\frac{\kappa}{N_v} x_1 x_2 = N_v \kappa c_1 c_2 \equiv N_v \lambda(c).$$

For unary reactions $S_1 \to *$,

$$\kappa x_1 = N_v \kappa c_1,$$

and for $2S_1 \to *$,

$$\frac{\kappa}{N_v} x_1 (x_1 - 1) = N_v \kappa c_1 (c_1 - N_v^{-1}) \approx N_v \kappa c_1^2. \tag{4.3}$$

© Springer International Publishing Switzerland 2015
D.F. Anderson, T.G. Kurtz, *Stochastic Analysis of Biochemical Systems*,
Mathematical Biosciences Institute Lecture Series 1,
DOI 10.1007/978-3-319-16895-1_4

Consequently, the equation for the species numbers in a system with m reactions becomes

$$X(t) = X(0) + \sum_{k=1}^{m} Y_k \left(N_v \int_0^t \lambda_k(C(s))ds\right)\zeta_k,$$

where $C(t) = N_v^{-1}X(t)$, and

$$C(t) = C(0) + \sum_{k=1}^{m} N_v^{-1}Y_k\left(N_v \int_0^t \lambda_k(C(s))ds\right)\zeta_k. \tag{4.4}$$

In the light of (4.3), we should write $\lambda_k^N(c)$ instead of $\lambda_k(c)$, but to keep the notation simpler, we will just write λ_k.

Even if the volume v is very small, N_v is still very large, and if we consider a sequence of equations

$$C^N(t) = C^N(0) + \sum_{k=1}^{m} N^{-1}Y_k\left(N \int_0^t \lambda_k(C^N(s))ds\right)\zeta_k, \tag{4.5}$$

where for $N = N_v$, $C^N = C$, C should be approximately equal to $\lim_{N\to\infty} C^N$, if the limit exists.

For simplicity, assume that $m < \infty$, and set $F(x) = \sum_{k=1}^{m} \lambda_k(x)\zeta_k$, so (4.5) becomes

$$C^N(t) = C^N(0) + M^N(t) + \int_0^t F(C^N(s))ds,$$

where

$$M^N(t) = \sum_{k=1}^{m} N^{-1}\tilde{Y}_k\left(N \int_0^t \lambda_k(C^N(s))ds\right)\zeta_k$$

and $\tilde{Y}_k(u) = Y_k(u) - u$. Assuming a local Lipschitz condition,

$$|F(x) - F(y)| \le K_a|x - y|, \quad |x|, |y| \le a, \tag{4.6}$$

set

$$x(t) = x(0) + \int_0^t F(x(s))ds, \tag{4.7}$$

and define $\gamma_a^N = \inf\{t : |C^N(t)| \vee |x(t)| \ge a\}$. Gronwall's inequality, Section A.7, gives

$$|C^N(t \wedge \gamma_a^N) - x(t \wedge \gamma_a^N)| \le (|C^N(0) - x(0)| + \sup_{s \le t \wedge \gamma_a^N} |M^N(s)|)e^{K_a t}.$$

Note that

$$M^N(t) = \sum_{k=1}^{m} N^{-1}\tilde{Y}_k\left(N \int_0^t \lambda_k(C^N(s))ds\right)\zeta_k$$

is a martingale with

$$E[|M^N(t \wedge \gamma_a^N)|^2] = \frac{1}{N}E[\int_0^{t \wedge \gamma_a^N} \sum_{k=1}^m \lambda_k(C^N(s))|\zeta_k|^2 ds],$$

and hence by Doob's inequality, Section A.2.1,

$$E[\sup_{s \leq t}|M^N(s \wedge \gamma_a^N)|^2] \leq \frac{4}{N}E[\int_0^{t \wedge \gamma_a^N} \sum_{k=1}^m \lambda_k(C^N(s))|\zeta_k|^2 ds]$$

$$\leq \frac{4t}{N}\sup_{|x| \leq a}\sum_{k=1}^m \lambda_k(x)|\zeta_k|^2.$$

Consequently, we have the following theorem. (See Chapter 11 of [19].)

Theorem 4.1. *Let C^N satisfy (4.5) and x satisfy (4.7). Suppose that for each $a > 0$, the Lipschitz condition (4.6) holds and that the solution of (4.7) exists for all time. (It is necessarily unique by the Lipschitz assumption.) If $C^N(0) \to x(0)$, then for each $\varepsilon > 0$ and each $t > 0$,*

$$\lim_{N \to \infty} P\{\sup_{s \leq t}|C^N(s) - x(s)| \geq \varepsilon\} = 0.$$

Theorem 4.1 is essentially a law of large numbers and could have been proved by a direct application of the law of large numbers for the Poisson processes which ensures

$$\lim_{N \to \infty}\sup_{u \leq u_0}|N^{-1}Y_k(Nu) - u| = 0 \quad a.s.$$

If one has a law of large numbers, then one should look for a central limit theorem, or in the case of stochastic processes, a functional central limit theorem, that is a limit theorem that captures the sample path behavior of the rescaled process.

With this goal in mind, consider

$$V^N(t) = \sqrt{N}(C^N(t) - x(t))$$

with C^N and x as above, and assume $V^N(0) \to V(0)$. Assuming that F is continuously differentiable, the stochastic equations give

$$V^N(t) = V^N(0) + \sum_{k=1}^m \frac{1}{\sqrt{N}}\tilde{Y}_k(N\int_0^t \lambda_k(C^N(s))ds)$$

$$+ \int_0^t \sqrt{N}(F(C^N(s)) - F(x(s)))ds$$

$$\approx V^N(0) + \sum_{k=1}^m \frac{1}{\sqrt{N}}\tilde{Y}_k(N\int_0^t \lambda_k(C^N(s))ds) + \int_0^t \nabla F(x(s)))V^N(s)ds.$$

By Theorem 4.1, we know that $\int_0^t \lambda_k(C^N(s))ds \to \int_0^t \lambda_k(x(s))ds$, and the standard central limit theorem implies that for $u \geq 0$,

$$W_k^N(u) = \frac{1}{\sqrt{N}}\tilde{Y}_k(Nu)$$

converges in distribution to a Gaussian random variable with mean zero and variance u. More precisely, the functional central limit theorem (Lemma A.6) holds; that is, $W_k^N \Rightarrow W_k$, where the W_k are independent standard Brownian motions.

Applying this functional convergence theorem for the W_k^N along with the continuous mapping theorem (Theorem A.5), we have $V^N \Rightarrow V$ satisfying

$$V(t) = V(0) + \sum_{k=1}^m W_k\left(\int_0^t \lambda_k(x(s))ds\right)\zeta_k + \int_0^t \nabla F(x(s))V(s)ds. \qquad (4.8)$$

Assuming that $V(0)$ is Gaussian, the linearity of (4.8) implies V is a Gaussian process. Taking expectations, we see that $E[V(t)]$ satisfies

$$E[V(t)] = E[V(0)] + \int_0^t \nabla F(x(s))E[V(s)]ds.$$

Setting

$$K(t) = \int_0^t \sum_k \zeta_k \zeta_k^T \lambda_k(x(s))ds,$$

one can also show that

$$E[V(t)V(t)^T] = E[V(0)V(0)^T] + K(t) + \int_0^t \nabla F(x(s))E[V(s)V(s)^T]ds$$

$$+ \int_0^t E[V(s)V(s)^T]\nabla F(x(s))^T ds,$$

and letting $\Gamma(t)$ denote the covariance matrix,

$$\Gamma(t) = \Gamma(0) + K(t) + \int_0^t \nabla F(x(s))\Gamma(s)ds + \int_0^t \Gamma(s)\nabla F(x(s))^T ds.$$

4.2 Models with multiple time-scales

Reaction networks modeled in cellular biology typically involve species numbers with orders of magnitude much smaller than the 10^{23} of Avogadro's number, and for some species, the numbers may be much too small to be reasonably modeled by continuous variables. Even for those species present in sufficient numbers (10^5 or even 10^3) to be modeled using continuous variables, expressing the species abundances as concentrations, as in the previous section, may not be appropriate, and there may be no normalization of species numbers that is appropriate for all species in the network. In addition, the rate constants in the models may vary over several orders of magnitude. Consequently, we want to explore the possibility of deriving simplified models, as in the previous section, by normalizing species numbers and rate constants in different ways.

Let $N_0 \gg 1$, where N_0 no longer has an interpretation in terms of Avogadro's number. For each species S_i, define the *normalized abundance* (or simply, the abundance) by

$$Z_i(t) = N_0^{-\alpha_i} X_i(t),$$

where $\alpha_i \geq 0$ should be selected so that $Z_i = O(1)$. Note that the abundance may be the species number ($\alpha_i = 0$) or the species concentration or something else.

Since the original rate constants, which we will denote by κ'_k, may also vary over several orders of magnitude, select β_k and κ_k so that $\kappa'_k = \kappa_k N_0^{\beta_k}$ and $\kappa_k = O(1)$. For binary reactions, for example, the rate function expressed in terms of the normalized abundances becomes

$$N_0^{\beta_k+\alpha_i+\alpha_j} \kappa_k z_i z_j = \kappa'_k x_i x_j.$$

Note that we can write $\alpha_i + \alpha_j = y_k \cdot \alpha$, where we recall that the kth reaction is $y_k \to y'_k$.

As before, we define a sequence of models satisfying

$$Z_i^N(t) = Z_i^N(0) + \sum^k N^{-\alpha_i} Y_k\left(\int_0^t N^{\beta_k+y_k\cdot\alpha}\lambda_k(Z^N(s))ds\right)\zeta_{ki},$$

where as before, $\zeta_k = y'_k - y_k$. Then the original model is $Z = Z^{N_0}$. The assumption that N_0 is "large" suggests attempting to derive approximate models by taking limits as $N \to \infty$. In these derivations, we want to exploit the possibility that the model has "multiple time-scales." To make clear what we mean by this terminology, consider a change of time variable replacing t by $N^\gamma t$, and define $Z^{N,\gamma}$ as the solution of the system

$$Z_i^{N,\gamma}(t) \equiv Z_i^N(N^\gamma t) = Z_i^N(0) + \sum^k N^{-\alpha_i} Y_k\left(\int_0^t N^{\gamma+\beta_k+y_k\cdot\alpha}\lambda_k(Z^{N,\gamma}(s))ds\right)\zeta_{ki}. \quad (4.9)$$

Assuming that the $Z_i^{N,\gamma}$ neither blow up nor converge uniformly to zero, each species has a natural time-scale γ_i determined by requiring

$$\alpha_i = \max\{\gamma_i + \beta_k + y_k \cdot \alpha : \zeta_{ki} \neq 0\}.$$

For $\gamma = \gamma_i$, none of the normalized reaction terms on the right of (4.9) should blow up, and at least one should be nontrivial. Since the γ_i need not all be the same, one may obtain different, nontrivial limiting models depending on the choice of time-scale γ.

This approach to deriving rescaled limits is explored in detail in [10] and [34]. In particular, [34] develops systematic approaches to the selection of the scaling exponents α_i and β_k. Here we simply derive limits for two examples.

4.2.1 Example: Derivation of the Michaelis-Menten equation

Classically, the Michaelis-Menten equation of enzyme kinetics is derived from the deterministic law of mass action under the assumption that the concentration of substrate is much larger than the concentration of enzyme. In [17], Darden derives the Michaelis-Menten equation from the Markov chain model under similar assumptions. We repeat that derivation using the stochastic equation representation we have developed here.

We consider the simplest system introduced in Section 2.1.3

$$S + E \rightleftharpoons SE \rightarrow P + E,$$

and we assume that the number of substrate molecules is $O(N)$ and the total number M of free enzyme and bound enzyme molecules is fixed and independent of N. Setting $Z_S^N(t) = N^{-1} X_S^N(t)$, we assume that the rate constants scale so that

$$X_E^N(t) = X_E(0) - Y_1\left(N \int_0^t \kappa_1 Z_S^N(s) X_E^N(s)ds\right) + Y_2\left(N\kappa_2 \int_0^t X_{SE}^N(s)ds\right)$$

$$+ Y_3\left(N\kappa_3 \int_0^t X_{SE}^N(s)ds\right)$$

$$Z_S^N(t) = Z_S^N(0) - N^{-1} Y_1\left(N \int_0^t \kappa_1 Z_S^N(s) X_E^N(s)ds\right)$$

$$+ N^{-1} Y_2\left(N\kappa_2 \int_0^t X_{SE}^N(s)ds\right),$$

where $M \equiv X_E^N(t) + X_{SE}^N(t)$ does not depend on t or N. In particular, we can substitute $M - X_E^N$ for X_{SE}^N. We assume that $X_E(0)$ does not depend on N and that $Z_S^N(0) \rightarrow Z_S(0) \equiv z_S(0) < \infty$.

For the equation written this way, the natural time-scale for the enzyme is $\gamma = -1$ and the natural time-scale for the substrate is $\gamma = 0$. It is easy to check that $(X_E^{N,-1}, Z_S^{N,-1})$ converges, as $N \rightarrow \infty$, to the solution of

$$X_E^{-1}(t) = X_E(0) - Y_1\left(\int_0^t \kappa_1 Z_S(s) X_E^{-1}(s)ds\right) + Y_2\left(\kappa_2 \int_0^t X_{SE}^{-1}(s)ds\right)$$

$$+ Y_3\left(\kappa_3 \int_0^t X_{SE}^{-1}(s)ds\right)$$

$$Z_S^{-1}(t) = Z_S(0).$$

For $\gamma = 0$ and hence $(X_E^{N,0}, Z_S^{N,0}) = (X_E^N, Z_S^N)$, the law of large numbers for the Poisson process implies that as $N \to \infty$,

$$Z_S^N(t) - Z_S^N(0) - \int_0^t (\kappa_2 X_{SE}^N(s) - \kappa_1 Z_S^N(s) X_E^N(s)) ds \to 0, \qquad (4.10)$$

almost surely, uniformly on bounded time intervals. Since $Z_S^N(0) \to z_S(0) < \infty$, the integrand in (4.10) is uniformly bounded, so at least along a subsequence, Z_S^N converges to a continuous function z_S. Similarly, dividing the equation for X_E^N by N, we have

$$\int_0^t \kappa_1 Z_S^N(s) X_E^N(s) ds - \kappa_2 \int_0^t (m - X_E^N(s)) ds - \kappa_3 \int_0^t (m - X_E^N(s)) ds$$
$$= \int_0^t (\kappa_1 Z_S^N(s) + \kappa_2 + \kappa_3) X_E^N(s) ds - (\kappa_2 + \kappa_3) mt \to 0,$$

which, by Problem 4.1(b), implies

$$\int_0^t X_E^N(s) ds \to \int_0^t \frac{m(\kappa_2 + \kappa_3)}{\kappa_1 z_S(s) + \kappa_2 + \kappa_3} ds,$$

at least along the subsequence described above.

Consequently

$$Z_S^N(t) = Z_S^N(0) - N^{-1} Y_1 \left(N \int_0^t \kappa_1 Z_S^N(s) X_E^N(s) ds \right)$$
$$+ N^{-1} Y_2 \left(N \kappa_2 \int_0^t X_{SE}^N(s) ds \right)$$
$$\approx Z_S^N(0) - \int_0^t \kappa_1 Z_S^N(s) X_E^N(s) ds + \kappa_2 \int_0^t X_{SE}^N(s) ds$$
$$\to z_S(0) - \kappa_3 \int_0^t \left(m - \frac{m(\kappa_2 + \kappa_3)}{\kappa_1 z_S(s) + \kappa_2 + \kappa_3} \right) ds$$
$$= z_S(0) - \int_0^t \frac{m \kappa_1 \kappa_3 z_S(s)}{\kappa_1 z_S(s) + \kappa_2 + \kappa_3} ds,$$

so $Z_S^N \to z_S$ satisfies

$$z_S(t) = z_S(0) - \int_0^t \frac{m \kappa_1 \kappa_3 z_S(s)}{\kappa_1 z_S(s) + \kappa_2 + \kappa_3} ds,$$

which is the Michaelis-Menten equation.

4.2.2 Example: Approximation of the virus model

We now consider the virus model introduced in Section 2.1.2 and studied earlier in [30, 45], and [10] using the rate constants from the original paper [45]. The system is

$$X_1(t) = X_1(0) + Y_1\left(\int_0^t X_3(s)ds\right) - Y_2\left(0.025\int_0^t X_1(s)ds\right)$$

$$-Y_6\left(7.5 \times 10^{-6}\int_0^t X_1(s)X_2(s)ds\right)$$

$$X_2(t) = X_2(0) + Y_3\left(\int_0^t 1000X_3(s)ds\right) - Y_5\left(2\int_0^t X_2(s)ds\right)$$

$$-Y_6\left(7.5 \times 10^{-6}\int_0^t X_1(s)X_2(s)ds\right)$$

$$X_3(t) = X_3(0) + Y_2\left(0.025\int_0^t X_1(s)ds\right) - Y_4\left(0.25\int_0^t X_3(s)ds\right)$$

Taking $N_0 = 1000$, we scale the rate constants so that

κ_1'	1	1
κ_2'	0.025	$2.5N_0^{-2/3}$
κ_3'	1000	N_0
κ_4'	0.25	0.25
κ_5'	2	2
κ_6'	7.5×10^{-6}	$0.75N_0^{-5/3}$

that is, we take $\beta_1 = \beta_4 = \beta_5 = 0$, $\beta_2 = -2/3$, $\beta_3 = 1$, and $\beta_6 = -5/3$. Scaling the species numbers, we take $\alpha_1 = 2/3$, $\alpha_2 = 1$, and $\alpha_3 = 0$.

With the scaled rate constants, the normalized equations become

$$Z_1^N(t) = Z_1^N(0) + N^{-2/3}Y_1\left(\int_0^t Z_3^N(s)ds\right) - N^{-2/3}Y_2\left(2.5\int_0^t Z_1^N(s)ds\right)$$

$$-N^{-2/3}Y_6\left(0.75\int_0^t Z_1^N(s)Z_2^N(s)ds\right)$$

$$Z_2^N(t) = Z_2^N(0) + N^{-1}Y_3\left(N\int_0^t Z_3^N(s)ds\right) - N^{-1}Y_5\left(N2\int_0^t Z_2^N(s)ds\right)$$

$$-N^{-1}Y_6\left(0.75\int_0^t Z_1^N(s)Z_2^N(s)ds\right)$$

$$Z_3^N(t) = Z_3^N(0) + Y_2\left(2.5\int_0^t Z_1^N(s)ds\right) - Y_4\left(0.25\int_0^t Z_3^N(s)ds\right).$$

With these choices of the α_i and β_k, we see that the natural time-scale for S_1 is $\gamma = 2/3$ and the natural time-scale for S_2 and S_3 is $\gamma = 0$.

Assuming $(Z_1^N(0), Z_2^N(0), Z_3^N(0)) \to (Z_1(0), Z_2(0), Z_3(0))$, taking the limit with $\gamma = 0$, we have

$$Z_1^0(t) = Z_1(0)$$

$$Z_2^0(t) = Z_2(0) + \int_0^t Z_3^0(s)ds - 2\int_0^t Z_2^0(s)ds$$

$$Z_3^0(t) = Z_3(0) + Y_2(2.5\int_0^t Z_1^0(s)ds) - Y_4(0.25\int_0^t Z_3^0(s)ds).$$

This system is an example of a *piecewise deterministic* or *hybrid* model, that is, one component is discrete and stochastic while the other is an ordinary differential equation with coefficients depending on the stochastic component.

For $\gamma = 2/3$, the system becomes

$$Z_1^{N,2/3}(t) = Z_1^{N,2/3}(0) + N^{-2/3}Y_1(N^{2/3}\int_0^t Z_3^{N,2/3}(s)ds)$$

$$-N^{-2/3}Y_2(N^{2/3}2.5\int_0^t Z_1^{N,2/3}(s)ds)$$

$$-N^{-2/3}Y_6(N^{2/3}0.75\int_0^t Z_1^{N,2/3}(s)Z_2^{N,2/3}(s)ds)$$

$$Z_2^{N,2/3}(t) = Z_2^{N,2/3}(0) + N^{-1}Y_3(N^{5/3}\int_0^t Z_3^{N,2/3}(s)ds)$$

$$-N^{-1}Y_5(N^{5/3}2\int_0^t Z_2^{N,2/3}(s)ds)$$

$$-N^{-1}Y_6(N^{2/3}0.75\int_0^t Z_1^{N,2/3}(s)Z_2^{N,2/3}(s)ds)$$

$$Z_3^{N,2/3}(t) = Z_3^{N,2/3}(0) + Y_2(N^{2/3}2.5\int_0^t Z_1^{N,2/3}(s)ds)$$

$$-Y_4(N^{2/3}0.25\int_0^t Z_3^{N,2/3}(s)ds).$$

As $N \to \infty$, dividing the equations for $Z_2^{N,2/3}$ and $Z_3^{N,2/3}$ by $N^{2/3}$ shows that

$$\int_0^t Z_3^{N,2/3}(s)ds - 2\int_0^t Z_2^{N,2/3}(s)ds \to 0$$

$$2.5\int_0^t Z_1^{N,2/3}(s)ds - 0.25\int_0^t Z_3^{N,2/3}(s)ds \to 0,$$

which together imply

$$\int_0^t Z_2^{N,2/3}(s)ds - 5\int_0^t tZ_1^{N,2/3}(s)ds \to 0.$$

Applying these limits in the equation for $Z_1^{N,2/3}$, estimates on the increments of $Z_1^{N,2/3}$ can be given that ensure that, at least along a subsequence, $Z_1^{N,2/3}$ converges to a continuous function z_1. Then applying Problem 4.1, we have

$$\int_0^t Z_1^{N,2/3}(s)Z_2^{N,2/3}(s)ds \to 5\int_0^t z_1(s)^2 ds$$

and the following theorem.

Theorem 4.2. *For each $\varepsilon > 0$ and $t > 0$,*

$$\lim_{N\to\infty} P\{\sup_{0\leq s\leq t} |Z_1^{N,2/3}(s) - z_1(s)| \geq \varepsilon\} = 0,$$

where z_1 is the solution of

$$z_1(t) = z_1(0) + \int_0^t 7.5z_1(s)ds) - \int_0^t 3.75z_1(s)^2 ds. \tag{4.11}$$

Problems

4.1. Let $x_n, x : [0,\infty) \to [0,\infty)$ and $z_n, z : [0,\infty) \to \mathbb{R}$ be measurable functions and z be continuous, and assume that for all $t > 0$, $\sup_{s\leq t} |z(s) - z_n(s)| \to 0$.

a) Suppose

$$\int_0^t x_n(s)ds \to \int_0^t x(s)ds < \infty,$$

for all $t > 0$. Show that

$$\int_0^t z_n(s)x_n(s)ds \to \int_0^t z(s)x(s)ds,$$

for all $t > 0$.
b) Suppose $T > 0$, $\inf_{s\leq T} z(s) > 0$, and

$$\int_0^t z_n(s)x_n(s)ds \to \int_0^t z(s)x(s)ds,$$

for all $0 < t \leq T$. Show that

$$\int_0^t x_n(s)ds \to \int_0^t x(s)ds,$$

for all $0 < t \leq T$.

4.2. Consider the following model for crystallization that was studied in [30]. The system involves four species and two reactions

$$2A \overset{\kappa'_1}{\to} B \qquad A + C \overset{\kappa'_2}{\to} D.$$

Rawlings and Haseltine assume $X_A(0) = 10^6$, $X_B(0) = 0$, $X_C(0) = 10$, and $\kappa'_1 = \kappa'_2 = 10^{-7}$.

a) Set up the system of stochastic equations for the model.
b) Select appropriate values for N_0 and the scaling exponents, and derive a simplified limiting model.

4.3. Consider an enzyme reaction of the form

$$S + E \rightleftharpoons SE \to P + E \qquad E \rightleftharpoons F + G \qquad \emptyset \to G \to \emptyset.$$

a) Are any linear combinations of species conserved? (Conservation relations reduce the number of equations you need to formulate the model.)
b) Assume that E, SE, and F are present only in small numbers and that S is present in much larger numbers. Assume that the reactions $SE \to S + E$, $SE \to P + E$, $E \rightleftharpoons F + G$ and $\emptyset \to G \to \emptyset$ are all fast, scale the model so that there are two time-scales, and identify the limiting models to the extent that you can.

4.4. Consider the network $\emptyset \to S_1 \to S_2 \to \emptyset$. Suppose the reaction $S_1 \to S_2$ is deterministic, that is, after a molecule of S_1 is created, it takes a deterministic amount of time to convert to a molecule of S_2. Assuming a "classical" scaling, prove a law of large numbers and a central limit theorem for the network.

4.5. Consider the network $\emptyset \to S_1 \rightleftharpoons S_2 \to \emptyset$. Use a continuous time Markov chain model, but now assume that the reversible reactions are much faster than the input and output reactions. Under appropriate scaling, prove a law of large numbers and a central limit theorem for the network.

Chapter 5
Numerical methods

5.1 Monte Carlo

When analyzing a stochastic model, one often wishes to approximate terms of the form $E[f(X)]$, where $f : D_E[0,\infty) \to \mathbb{R}$ is a scalar-valued functional of a path which gives a measurement of interest. Examples of functionals f include

- $f(X(T)) = X_i(T)$, yielding estimates for mean values at a specific time,
- $f(X) = t^{-1} \int_0^t g(X(s))ds$, yielding time averages,
- $f(X(T)) = X_i(T)X_j(T)$, yielding covariances,
- $f(X) = 1_{\{X(T)\in B\}}$, yielding probabilities,

though this list is far from exhaustive.

Suppose for the time being that we can generate realizations of X exactly via a computer, and that $E[|f(X)|] < \infty$. The strong law of large numbers tells us that if $X_{[1]}, X_{[2]}, \ldots$, are independent realizations of X then with a probability of one

$$\lim_{n \to \infty} \frac{f(X_{[1]}) + \cdots + f(X_{[n]})}{n} = E[f(X)].$$

This of course implies the beginnings of a strategy: generate large numbers of independent realizations $\{X_{[i]}\}_{i=1}^n$, and use the approximation

$$E[f(X)] \approx \frac{1}{n} \sum_{i=1}^{n} f(X_{[i]}). \tag{5.1}$$

We are now immediately confronted with the following question: How good is the approximation (5.1)? This question can be answered via the central limit theorem, which for completeness we restate here. We point the reader to Appendix A.4 for more details related to convergence of distribution.

© Springer International Publishing Switzerland 2015
D.F. Anderson, T.G. Kurtz, *Stochastic Analysis of Biochemical Systems*,
Mathematical Biosciences Institute Lecture Series 1,
DOI 10.1007/978-3-319-16895-1_5

Theorem 5.1 (Central limit theorem). *Let $X_{[1]}, X_{[2]}, \ldots$ be a sequence of independent and identically distributed $\mathbb{R}$-valued random variables with $E[X_{[i]}] = \mu \in (-\infty, \infty)$ and $Var(X_{[i]}) = \sigma^2 < \infty$. If $S_n = X_{[1]} + \cdots + X_{[n]}$, then*

$$\frac{S_n - n\mu}{\sigma n^{1/2}} \Rightarrow Z, \text{ as } n \to \infty,$$

where Z is a standard normal.

The conclusion of Theorem 5.1 says that for any $x \in \mathbb{R}$, we have

$$\lim_{n \to \infty} P\left\{\frac{S_n - n\mu}{\sigma n^{1/2}} \leq x\right\} = \frac{1}{\sqrt{2\pi}} \int_{-\infty}^{x} e^{-s^2/2} ds.$$

Returning to our estimator $n^{-1} \sum_{i=1}^{n} f(X_{[i]})$ in (5.1), the central limit theorem implies that for $\varepsilon > 0$,

$$P\left\{\left|\frac{1}{n}\sum_{i=1}^{n} f(X_{[i]}) - E[f(X)]\right| \leq \varepsilon\right\} \approx P\left\{-\frac{\sqrt{n}\varepsilon}{\sigma} \leq Z \leq \frac{\sqrt{n}\varepsilon}{\sigma}\right\},$$

where Z is a standard normal. Note that σ is most likely unknown to us, so letting s_n^2 be the usual unbiased estimator for the variance of a population, we take

$$P\left\{\left|\frac{1}{n}\sum_{i=1}^{n} f(X_{[i]}) - E[f(X)]\right| \leq \varepsilon\right\} \approx P\left\{-\frac{\sqrt{n}\varepsilon}{s_n} \leq Z \leq \frac{\sqrt{n}\varepsilon}{s_n}\right\}. \tag{5.2}$$

Letting $\hat{\mu}_n = n^{-1} \sum_{i=1}^{n} f(X_{[i]})$ we conclude that the probability that the true, unknown, value $E[f(X)]$ is in the interval $(\hat{\mu}_n - \varepsilon, \hat{\mu}_n + \varepsilon)$ is approximately given by the right-hand side of the above equation. The interval $(\hat{\mu}_n - \varepsilon, \hat{\mu}_n + \varepsilon)$ is termed a *confidence interval*.

Example 5.2. Suppose that after $n = 1,000$ trials we observe $\hat{\mu}_n = 13.45$ and $s_n^2 = 3.26$. Then, since $2\Phi(\sqrt{1000} \times 0.119/\sqrt{3.26}) - 1 \approx 0.95$, where Φ is the cumulative distribution function for a standard normal random variable, the 95% confidence interval is $(13.45 - 0.1119, 13.45 + 0.1119)$. $\triangle$

5.2 Generating random variables: Transformations of uniforms

In order to generate realizations of stochastic processes on a computer, it is crucial that we are able to generate random variables efficiently. There are now sophisticated algorithms that are found in almost all mathematics software packages that generate pseudo-uniform random variables defined on the interval $[0, 1]$. A discussion of the quality of these pseudo-random variables is not appropriate for this text, and we will simply assume here that the uniforms so generated are, in fact, actual independent uniform random variables.

Supposing now that we may generate uniform random variables, we consider how to transform them to get random variables with other distributions. First, note that if $g : \mathbb{R} \to \mathbb{R}$ is a nondecreasing, right continuous function and we define

$$g^{-1}(t) = \inf\{x : g(x) \geq t\} = \min\{x : g(x) \geq t\},$$

then $\{x : g(x) \geq t\} = \{x : x \geq g^{-1}(t)\}$, and hence, if X is a random variable with cumulative distribution function F_X, then for $t \in \mathbb{R}$

$$P\{g(X) < t\} = P\{X < g^{-1}(t)\} = F_X(g^{-1}(t)-).$$

Similarly, if g is right continuous and nonincreasing and

$$g^{-1}(t) = \inf\{x : g(x) \leq t\} = \min\{x : g(x) \leq t\},$$

then $\{x : g(x) \leq t\} = \{x : x \geq g^{-1}(t)\}$, and hence,

$$P\{g(X) \leq t\} = P\{X \geq g^{-1}(t)\} = 1 - F_X(g^{-1}(t)-).$$

Example 5.3. Let $U \sim \text{uniform}[0,1]$, $\lambda > 0$, and let $g : (0,1] \to \mathbb{R}$ be

$$g(x) = -\ln(x)/\lambda = \ln(1/x)/\lambda.$$

Then $g(U)$ is exponentially distributed with parameter $\lambda > 0$.

First note that for $t < 0$, we obviously have that $P\{g(U) \leq t\} = 0$. Next, when $t \geq 0$ we have

$$P\{g(U) \leq t\} = P\{\ln(1/U)/\lambda \leq t\} = P\left\{U \geq e^{-\lambda t}\right\} = 1 - e^{-\lambda t}.$$

$\triangle$

We will also need to generate discrete random variables.

Example 5.4. Suppose that a random variable X takes values in the discrete set $K = \{k_1, k_2, \ldots\}$ with associated probabilities $P\{X = k_i\} = p_i$, where $\sum_i p_i = 1$. Note that K could be finite or countably infinite. Define $g : [0,1] \to K$ in the following manner: $g(x) = k_j$ if and only if

$$\sum_{i=1}^{j-1} p_i < x \leq \sum_{i=1}^{j} p_i,$$

where we define sums of the form $\sum_{i=1}^{0} a_i$ to be zero. If $U \sim \text{uniform}[0,1]$, then $g(U)$ has the same distribution as X.

To see this, simply note that for $U \sim \text{uniform}[0,1]$,

$$P\{g(U) = k_j\} = P\left\{\sum_{i=1}^{j-1} p_i < U \leq \sum_{i=1}^{j} p_i\right\} = p_j.$$

$\triangle$

5.3 Exact simulation methods

We provide two methods for the generation of statistically exact sample paths of stochastic models of biochemical processes. In Section 5.3.1 we provide the algorithm which simulates the embedded discrete time Markov chain concurrently with the exponential holding times. This algorithm is commonly called "Gillespie's algorithm" in the current context. In Section 5.3.2 we provide the next reaction method which simulates the random time change representation (2.3).

5.3.1 Embedded discrete time Markov chains and the stochastic simulation algorithm

Let $X(t)$ be a continuous time Markov chain. Supposing that $\{t_j\}_{j=0}^{\infty}$ are the transition times of X, we define $Z_j = X(t_j)$ for all $j \geq 0$. The Markov chain Z is called the *embedded discrete time Markov chain* associated with $X(t)$. Conversely, note that the process $X(t)$ is completely determined by the embedded chain Z_j and the (exponentially distributed) holding times $\{t_{j+1} - t_j\}_{j=0}^{\infty}$.

The stochastic simulation algorithm, or *Gillespie's algorithm*, proceeds by simulating the embedded discrete time Markov chain concurrently with the exponentially distributed holding times. Specifically, we simulate the discrete time Markov chain with transition probabilities

$$p_{xy} = \begin{cases} \frac{\lambda_k(x)}{\sum_\ell \lambda_\ell(x)} & , \quad \text{if } y = x + \zeta_k \\ 0 & , \qquad \text{else} \end{cases}$$

in order to account for the transitions in the model, and simulate an exponential random variable with a parameter of $\sum_k \lambda_k(x)$ to account for the holding time in state x. To generate these random variables, we simply use the methods detailed in Section 5.2.

In the algorithm below, all uniform random variables are assumed to be mutually independent. Also, note that the algorithm below only provides the state of the system at the jump times $\{t_j\}_{j=0}^{\infty}$. The process is, of course, defined for all t by taking $X(t) = X(t_j)$ for all $t \in [t_j, t_{j+1})$. See [25, 26] for one development of the following algorithm.

(Stochastic simulation algorithm / Gillespie's Algorithm)

Initialize: Given: a chemical reaction network with intensity functions λ_k and jump directions ζ_k, $k = 1, \ldots, R$, and initial condition x_0. Set $j = 0$, $t_0 = 0$, and $X(t_0) = x_0 \in \mathbb{Z}_{\geq 0}^n$.

Repeat the following steps.

1. For all $k \in \{1, \ldots, R\}$, calculate $\lambda_k(X(t_j))$.
2. Set $\lambda_0(X(t_j)) = \sum_{k=1}^{R} \lambda_k(X(t_j))$.
3. Generate two independent uniform(0,1) random numbers r_{j1} and r_{j2}.
4. Set $\Delta = \ln(1/r_{j1})/\lambda_0(X(t_j))$ and $t_{j+1} = t_j + \Delta$.
5. Find $\mu \in [1, \ldots, R]$ such that

$$\frac{1}{\lambda_0(X(t_j))} \sum_{k=1}^{\mu-1} \lambda_k(X(t_j)) < r_{j2} \leq \frac{1}{\lambda_0(X(t_j))} \sum_{k=1}^{\mu} \lambda_k(X(t_j)),$$

   and set $X(t_{j+1}) = X(t_j) + \zeta_\mu$.
6. Set $j \leftarrow j + 1$.

Note that the above algorithm uses two random numbers per step. The first is used to find *when* the next reaction occurs and the second is used to determine *which* reaction occurs at that time.

Sample MATLAB code that implements the stochastic simulation algorithm on the model found in Example 2.5 can be found online as supplementary material.

5.3.2 The next reaction method

Simulation of the random time change representation (2.3) is called the next reaction method in the present context, see [1, 23]. Before presenting the algorithm, we provide a motivating example.

Example 5.5. Consider the reaction network

$$S \underset{\kappa_2}{\overset{\kappa_1}{\rightleftharpoons}} 2S,$$

which has associated stochastic equation

$$X(t) = X(0) + Y_1\left(\int_0^t \kappa_1 X(s)ds\right) - Y_2\left(\int_0^t \kappa_2 X(s)(X(s)-1)ds\right),$$

where $X(t)$ provides the count of S at time t. We now think about how to simulate this process forward in time with the main observation that X is simply a function of the Poisson processes Y_1 and Y_2. Further, as Poisson processes can be constructed via their exponential holding times, we suppose that $\{e_{11}, e_{12}, \ldots\}$ and $\{e_{21}, e_{22}, \ldots\}$ are independent unit exponential random variables and suppose further that for $k \in$

$\{1,2\}$ the jump times of the process Y_k are given by $e_{k1}, e_{k1} + e_{k2}, \ldots$. Finally, for all $t \geq 0$ we define $T_1(t) = \int_0^t \kappa_1 X(s) ds$, $T_2(t) = \int_0^t \kappa_2 X(s)(X(s) - 1) ds$ and for each $k \in \{1,2\}$,

$$P_k(t) = \inf\{s > T_k(t) : Y_k(s) - Y_k(T_k(t)) > 0\}.$$

That is, $P_k(t)$ simply gives the time of the next jump of Y_k after time $T_k(t)$.

In order to simulate the above representation, we now simply note that $P_k(0) = e_{k1}$ for each $k \in \{1,2\}$, and that the process $X(t)$ is constant up to time

$$\Delta = \min\left\{ \frac{P_1(0) - T_1(0)}{\lambda_1(X(0))}, \frac{P_2(0) - T_2(0)}{\lambda_2(X(0))} \right\}, \tag{5.3}$$

since no counting process will change until one of $\int_0^t \lambda_k(X(s)) ds = t \lambda_k(X(0))$ hits $e_{k1} = P_k(0)$, and since $T_k(0) = 0$ for each k.

Now suppose that the minimum was achieved by reaction 2 at time Δ. In this case, we set $t_1 = \Delta$, $X(t_1) = X(0) - 1$ (since it was reaction 2 that happened), and update

$$T_k(t_1) = \int_0^{t_1} \lambda_k(X(s)) ds = t_1 \cdot \lambda_k(X(0)), \quad \text{for each } k \in \{1,2\}.$$

Next, we note that $P_1(t_1) = P_1(0)$ (since that point was not reached), but that

$$P_2(t_1) = \inf\{s > T_2(t_1) : Y_2(s) > Y_2(T_2(t_1))\}.$$

Note that, in fact, $T_2(t_1) = e_{21}$ (since we have now hit that point) and $P_2(t_1) = e_{21} + e_{22}$. Continuing, we may now conclude that the process is constant from t_1 until $t_1 + \Delta$ for Δ satisfying

$$\Delta = \min\left\{ \frac{P_1(t_1) - T_1(t_1)}{\lambda_1(X(t_1))}, \frac{P_2(t_1) - T_2(t_1)}{\lambda_2(X(t_1))} \right\}.$$

The algorithm now proceeds by continually updating T_k and P_k as above and asking which reaction will be the next to achieve the minimum in (5.3).   △

In the algorithm below, we consider a chemical reaction network with intensity functions λ_k and jump directions ζ_k, $k = 1, \ldots, R$. For each k and all $t \geq 0$ we let

$$T_k(t) = \int_0^t \lambda_k(X(s)) ds, \quad \text{and} \quad P_k(t) = \inf\{s > T_k(t) : Y_k(s) > Y_k(T_k(t))\}.$$

All generated uniform random variables are assumed to be mutually independent.

(Next Reaction Method)

Initialize: Given: a chemical reaction network with intensity functions λ_k and jump directions ζ_k, $k = 1, \ldots, R$, and initial condition x_0. Set $j = 0$, $t_0 = 0$, and $X(t_0) = x_0 \in \mathbb{Z}^n_{\geq 0}$. For each k, set $T_k(t_0) = 0$ and $P_k(t_0) = \ln(1/r_{k0})$, where r_{k0} are independent uniform$[0, 1]$ random variables.

Repeat the following steps.

1. For all $k \in \{1, \ldots, R\}$, calculate $\lambda_k(X(t_j))$.
2. For each k, set

$$\Delta t_k = \begin{cases} (P_k(t_j) - T_k(t_j))/\lambda_k(X(t_j)), & \text{if } \lambda_k(X(t_j)) \neq 0 \\ \infty, & \text{if } \lambda_k(X(t_j)) = 0 \end{cases}.$$

3. Set $\Delta = \min_k\{\Delta t_k\}$, and let μ be the index where the minimum is achieved.
4. Set $t_{j+1} = t_j + \Delta$ and $X(t_{j+1}) = X(t_j) + \zeta_\mu$.
5. For each $k \in \{1, \ldots, M\}$, set $T_k(t_{j+1}) = T_k(t_j) + \lambda_k(X(t_j)) \cdot \Delta$.
6. Set $P_\mu(t_{j+1}) = P_\mu(t_j) + \ln(1/r_j)$, where r_j is a uniform$(0,1)$ random variable, and for $k \neq \mu$ set $P_k(t_{j+1}) = P_k(t_j)$.
7. Set $j \leftarrow j + 1$.

Note that after initialization the Next Reaction Method only demands one random number to be generated per step.

Sample MATLAB code that implements the next reaction method on the model found in Example 2.5 can be found online as supplementary material.

5.3.2.1 Time dependent intensity functions

Due to changes in temperature, volume, or voltage (in the case of a model of a neuronal system such as Morris-Lecar or Hodgkin Huxley), the rate parameters of a system may be functions of time. That is, we may have $\lambda_k(t) = \lambda_k(X(t), t)$, and the stochastic equations for the model become

$$X(t) = X(0) + \sum_k Y_k\left(\int_0^t \lambda_k(X(s), s)ds\right)\zeta_k. \tag{5.4}$$

The next reaction method as presented above is easily modified to incorporate this time dependence. The only step that would change is step 2., which becomes:

2. For each k, find Δt_k satisfying

$$\int_t^{t+\Delta t_k} \lambda_k(X(s), s)ds = P_k(t_j) - T_k(t_j).$$

Note, in particular, that the integral ranges from t to $t + \Delta t_k$.

5.4 Approximate simulation with Euler's method / τ-leaping

Each of the above statistically exact algorithms is based on an exact accounting of all the reaction events that take place. In highly complex systems such exact algorithms become computationally burdensome since the number of computations scales linearly with the number of reaction events. Conditioned on the current state $X(t)$, the holding time is exponentially distributed with mean $\Delta_t = 1/\sum_k \lambda_k(X(t))$. If $\Delta_t \ll 1$ for all t, then the time needed to produce a single sample path can be prohibitive. This problem is greatly amplified by the fact that these simulations are usually paired with Monte Carlo techniques that require the generation of many sample paths.

To address the problem that Δ_t may be prohibitively small, approximate algorithms, and notably the class of algorithms termed "tau-leaping" methods introduced by Gillespie [28], have been developed with the explicit aim of greatly lowering the computational complexity of each path simulation while attempting to control the bias (see, for a small subset of this rather large literature, [2, 5, 7, 28, 38, 44]). Tau-leaping is essentially Euler's method in that for some time discretization parameter $h > 0$, the process generated by tau-leaping can be represented via the equation

$$Z(t) = Z(0) + \sum_k Y_k \Big(\int_0^t \lambda_k(Z \circ \eta(s)) ds \Big) \zeta_k, \tag{5.5}$$

where $\eta(s) = \lfloor s/h \rfloor \cdot h$ fixes the state of the process at the left endpoint of the discretization. Note that if $\{t_j\}$ is the discretization of $[0, \infty)$ with $t_{j+1} - t_j = h$ for all $j \geq 0$, then we have

$$\int_0^{t_{j+1}} \lambda_k(Z \circ \eta(s)) ds = \sum_{i=0}^{j} \lambda_k(Z(t_i))(t_{i+1} - t_i),$$

which explains why tau-leaping can be understood as an Euler method. While the most straightforward version of tau-leaping is presented below, more general procedures (i) determine a new h at each step [2, 14, 27], and (ii) guarantee that trajectories can never leave the positive orthant [2, 13, 15, 46]. Implicit [43], midpoint [7], and Runge-Kutta methods [12] have also been developed.

The following algorithm simulates the process (5.5) at the discretization time points $\{t_j\}$, and, unlike the exact methods, does not provide values of Z between t_j and t_{j+1}. Below, for $x \geq 0$ we will write Poisson(x) to denote a sample from the Poisson distribution with parameter x, with all such samples being independent of each other and of all other sources of randomness used.

(Euler tau-leaping)

Initialize: Given: a chemical reaction network with intensity functions λ_k and jump directions ζ_k, $k = 1, \ldots, R$, and initial condition x_0. Fix $h > 0$ and set $j = 0$, $t_0 = 0$, and $X(t_0) = x_0 \in \mathbb{Z}_{\geq 0}^n$.

Repeat the following steps.

1. Set $t_{j+1} = t_j + h$.
2. For each k, let $\Lambda_k = \text{Poisson}(\lambda_k(Z_h(t_j))h)$.
3. Set $Z_h(t_{j+1}) = Z_h(t_j) + \sum_k \Lambda_k \zeta_k$.
4. Set $j \leftarrow j + 1$.

5.5 Monte Carlo and multi-level Monte Carlo

The exact and approximate methods outlined so far in this chapter can be combined in interesting ways in order to efficiently approximate expectations.

5.5.1 Computational complexity for Monte Carlo

We return to the context of Section 5.1. We let X be a realization of a chemical process and consider the computational complexity, or number of computations, required by a computer to approximate $E[f(X)]$ via Monte Carlo to $O(\varepsilon)$ accuracy in the sense of confidence intervals. Equation (5.2) implies that to achieve such an order of accuracy, we must get the standard deviation of our estimator,

$$\hat{\mu}_n = \frac{1}{n} \sum_{i=1}^{n} f(X_{[i]}),$$

to be $O(\varepsilon)$. Since

$$\text{Var}(\hat{\mu}_n) = \frac{1}{n} \text{Var}(f(X)),$$

we see that the number of independent sample paths required is $O(\varepsilon^{-2}\text{Var}(f(X)))$. If we let $\overline{N} > 0$ be the order of magnitude of the number of computations needed to produce a single sample path using an exact algorithm, then the total computational complexity becomes $O(\overline{N}\varepsilon^{-2}\text{Var}(f(X)))$.

Since for many models of interest the term $\overline{N}$ is very large (e.g., for some models, simply generating a few dozen paths is prohibitive), approximate algorithms, and notably tau-leaping methods (5.5) are often employed for the Monte Carlo computation. In this case, we let

$$\hat{\mu}_n^Z = \frac{1}{n} \sum_{i=1}^{n} f(Z_{\ell,[i]}),$$

where $Z_{\ell,[i]}$ is the ith independent approximate path constructed with a time discretization parameter of h_ℓ. Note that

$$E[f(X)] - \hat{\mu}_n^Z = (E[f(X)] - E[f(Z_\ell)]) + (E[f(Z_\ell)] - \hat{\mu}_n^Z). \qquad (5.6)$$

Supposing that we are using a weakly first order numerical method so that $E[f(X)] - E[f(Z_\ell)] = O(h_\ell)$, and still supposing that we want to approximate $E[f(X)]$ to an accuracy of $\varepsilon > 0$ with a given confidence, then, (*i*) we must choose $h_\ell = O(\varepsilon)$ so the first term on the right of (5.6), the bias, is $O(\varepsilon)$, and (*ii*) we must generate $O(\varepsilon^{-2}\mathrm{Var}(f(Z_\ell)))$ paths so the Monte Carlo estimator for the second term on the right of (5.6), the statistical error, has a standard deviation of $O(\varepsilon)$. We then find a total computational complexity of $O(\varepsilon^{-3}\mathrm{Var}(f(Z_\ell)))$. This computational complexity can be less than that required by an exact algorithm, $O\left(\varepsilon^{-2}\overline{N}\,\mathrm{Var}(f(X))\right)$, when $\varepsilon^{-1} \ll \overline{N}$.

Note that in both cases above the variance of the random variable being estimated was a critical parameter in the computational complexity of the problem. Variance reduction methods can sometimes yield large savings in computational time, with one such method briefly introduced in the next section.

5.5.2 Multi-level Monte Carlo (MLMC)

In the diffusive setting, in which the noise enters the system via Brownian motions as opposed to the Poisson processes of the current context, the multi-level Monte Carlo (MLMC) approach of Giles [24], with related earlier work by Heinrich [31], has the remarkable property of lowering the standard $O(\varepsilon^{-3})$ cost of computing an $O(\varepsilon)$ accurate Monte Carlo estimate of $E[f(X)]$ down to $O(\varepsilon^{-2}\log(\varepsilon)^2)$. Here, we are assuming that a weak order one and strong order $1/2$ discretization method, such as Euler–Maruyama, is used.

We can motivate the general multi-level philosophy by observing that we are not obliged to use the same discretization level, h, for every sample path of an approximate algorithm. Given that a smaller h incurs a higher cost, it may be beneficial to mix together samples at different h-resolutions. We show how to do this in the present setting by following the work [4, 9].

The MLMC estimator of [4] is built in the following manner. For a fixed integer M, and $\ell \in \{\ell_0, \ell_0 + 1, \ldots, L\}$, where both ℓ_0 and L depend upon the model and path-wise simulation method being used, let $h_\ell = TM^{-\ell}$, where T is our terminal time. The value of M can be specified by the user, but reasonable choices for M are $M \in \{2, 3, \ldots, 7\}$. Note that

$$E[f(X)] = E[f(X) - f(Z_L)] + \sum_{\ell=\ell_0+1}^{L} E[f(Z_\ell) - f(Z_{\ell-1})] + E[f(Z_{\ell_0})], \qquad (5.7)$$

where the telescoping sum is the key feature. At this point, it certainly appears we have made things worse, as now we need estimators for each of the expectations above. Continuing on, we define independent estimators for the multiple terms by

$$\hat{Q}_E = \frac{1}{n_E} \sum_{i=1}^{n_E} (f(X_{[i]}) - f(Z_{L,[i]})),$$

$$\hat{Q}_{\ell_0} = \frac{1}{n_0} \sum_{i=1}^{n_0} f(Z_{\ell_0,[i]}), \tag{5.8}$$

$$\hat{Q}_\ell = \frac{1}{n_\ell} \sum_{i=1}^{n_\ell} (f(Z_{\ell,[i]}) - f(Z_{\ell-1,[i]})),$$

where $\ell \in \{\ell_0 + 1, \ldots, L\}$, and note that

$$\hat{Q} = \hat{Q}_E + \sum_{\ell=\ell_0}^{L} \hat{Q}_\ell \tag{5.9}$$

is an *unbiased* estimator for $E[f(X)]$. The above observations are not useful unless we can successfully *couple* the process (X, Z_L) and $(Z_\ell, Z_{\ell-1})$ together in a way that significantly reduces the variance of the estimator $\hat{Q}_{(\cdot)}$ at each level. Defining

$$\Lambda_k(X, Z_L, s) = \min\{\lambda_k(X(s)), \lambda_k(Z_L \circ \eta_L(s))\}$$

to be the minimum of the respective intensity functions, the coupling introduced in [4] for the processes X and Z_L is

$$X(t) = X(0) + \sum_k \left[Y_{k,1}\left(\int_0^t \Lambda_k(X, Z_L, s) ds \right) \right.$$
$$\left. + Y_{k,2}\left(\int_0^t \{\lambda_k(X(s)) - \Lambda_k(X, Z_L, s)\} ds \right) \right] \zeta_k$$
$$Z_L(t) = Z_L(0) + \sum_k \left[Y_{k,1}\left(\int_0^t \Lambda_k(X, Z_L, s) ds \right) \right. \tag{5.10}$$
$$\left. + Y_{k,3}\left(\int_0^t \{\lambda_k(Z_L \circ \eta_L(s)) - \Lambda_k(X, Z_L, s)\} ds \right) \right] \zeta_k,$$

which forces the marginal processes to jump together through the counting processes $Y_{k,1}(\cdot)$. There is a similar coupling for the processes $(Z_\ell, Z_{\ell-1})$; see [4]. Note that arguments similar to those in Problem 1.4 explain why these processes have the correct marginal distributions. Numerical simulation of the coupling (5.10) can be carried out via any exact algorithm, whereas the coupled processes $(Z_\ell, Z_{\ell-1})$ can be simulated via a version of tau-leaping. Numerical examples in [4] demonstrated speedups by factors of over 100, with no loss in accuracy.

We turn to the question of *why* the multi-level Monte Carlo estimator reduces to $O(\varepsilon^{-2} \ln(\varepsilon)^2)$ the computational complexity of approximating $E[f(X)]$ to an order of accuracy of $O(\varepsilon)$. We let X denote our generic exact process and for $\ell \in \{0,1,\ldots,L\}$, where L is to be determined, we let Z_ℓ be an approximate process constructed with a time discretization of $h_\ell = T/M^\ell$. The function f still gives our statistic of interest. For simplicity, we assume that we cannot generate realizations of the process X, which is common in the diffusive setting. If the processes X can be simulated, as in the case considered here, we simply add the estimator $\hat{Q}_E$ to the construction below and the analysis changes somewhat [4]. We make the following running assumptions on our processes.

Running Assumptions.

1.) $E[|f(X(t))|] < \infty$, for all $t \geq 0$.
2.) There is a constant $C_1 > 0$ for which $|E[f(X(t))] - E[f(Z_\ell(t))]| < C_1 h_\ell$.
3.) There is a constant $C_2 > 0$ for which $\text{Var}(f(X) - f(Z_\ell)) < C_2 h_\ell$.

The assumptions are satisfied, for example, if the numerical method is order 1 accurate in a weak sense, and order 1/2 accurate in a strong sense. For $\ell \in \{0,1,\ldots,L\}$ we let $\hat{Q}_\ell$ be as in (5.8) and let $\hat{Q} = \sum_{\ell=0}^{L} \hat{Q}_\ell$. Note that $\hat{Q}$ is an unbiased estimator for $E[f(Z_L)]$ and

$$E[f(X)] - \hat{Q} = (E[f(X)] - E[f(Z_L)]) + (E[f(Z_L)] - \hat{Q}).$$

In order to provide an estimate of accuracy $\varepsilon > 0$ in terms of confidence intervals it is sufficient to ensure that the bias, $E[f(X)] - E[f(Z_L)]$, is $O(\varepsilon)$ and that the standard deviation of $\hat{Q}$ is of order ε; that is, that the variance of $\hat{Q}$ is of order ε^2.

Handling the bias is straightforward. Choose $L = O(|\ln(\varepsilon)|)$ so that $h_L = O(\varepsilon)$ and so

$$|E[f(X)] - E[f(Z_L)]| = O(\varepsilon),$$

where we used the running assumption 2.).

Turning to the statistical error, we note that since the estimators are built using independent paths

$$\text{Var}(\hat{Q}) = \text{Var}\left(\sum_{\ell=0}^{L} \hat{Q}_\ell\right) = \sum_{\ell=0}^{L} \text{Var}(\hat{Q}_\ell).$$

Further, by the running assumption 3.), for $\ell \in \{1,\ldots,L\}$ we have

$$\text{Var}(\hat{Q}_\ell) = \frac{1}{n_\ell}\text{Var}(f(Z_\ell) - f(Z_{\ell-1})) \leq \frac{1}{n_\ell}C_2 h_\ell.$$

Since $\text{Var}(\hat{Q}_0) = O(n_0^{-1})$ we have that for some constant $C > 0$,

$$\text{Var}(\hat{Q}) \leq C\left(\sum_{\ell=1}^{L} \frac{1}{n_\ell}h_\ell + \frac{1}{n_0}\right).$$

Choosing $n_0 = O(\varepsilon^{-2})$ and $n_\ell = O(\varepsilon^{-2}h_\ell L)$, a straightforward calculation shows that

$$\mathsf{Var}(\hat{Q}) = O(\varepsilon^2),$$

as desired. We now compute the computational complexity of generating the above described sequence of estimators. First, the cost of generating $\hat{Q}_0$ is of order

$$\text{\# required paths} \times \text{\# steps per path} = \varepsilon^{-2} \times 1 = \varepsilon^{-2}.$$

Next, the computational complexity of generating $\hat{Q}_\ell$ for $\ell \in \{1, 2, \ldots, L\}$ is of order

$$\text{\# required paths} \times \text{\# steps per path} = \varepsilon^{-2}h_\ell L \times h_\ell^{-1} = \varepsilon^{-2}L.$$

Hence, the total computational complexity is of order

$$\sum_{\ell=1}^{L} \varepsilon^{-2}L + \varepsilon^{-2} = \varepsilon^{-2}(L^2 + 1).$$

Recalling that $L = O(|\ln(\varepsilon)|)$ completes the argument.

We note that the gains in computational efficiency come about for two reasons. First, a coordinated sequence of simulations is being done, with nested step-sizes, and the simulations with larger step-size are much cheaper than those with very fine step-sizes. Second, while we do still require the generation of paths with fine step-sizes, the variance of $f(Z_\ell) - f(Z_{\ell-1})$ will be small, thereby requiring significantly fewer of these expensive paths in the estimation of $\hat{Q}_\ell$.

The reduction in computational complexity predicted by the analysis above is observed in examples.

Problems

5.1. Find a 99% confidence interval for the experiment detailed in Example 5.2. Find both a 95% and 99% confidence interval for the case $n = 10,000$, $\hat{\mu}_n = 13.45$ and $s_n^2 = 3.26$.

5.2. Using Gillespie's algorithm, simulate and plot a single trajectory of

$$S_1 \overset{2}{\underset{1}{\rightleftharpoons}} S_2,$$

up to time $T = 10$ under the assumption that $S_1(0) = 15$ and $S_2(0) = 0$. Find a 95% confidence interval for $E[X_1(10)]$ using 1,000 independent simulations of the process.

5.3. Using the next reaction method, simulate and plot a single trajectory of the model in Example 2.3 with $\kappa_1 = 200$, $\kappa_2 = 10$, $d_M = 25$, $d_p = 1$, an initial condition of 1 gene, 10 mRNA, and 50 protein molecules, and a terminal time of $T = 8$. Note that you are asked to produce a plot similar to that of Figure 2.1. Find a 95% confidence interval for $E[X_{\text{protein}}(8)]$ using 1,000 independent simulations of the process.

5.4. Using Euler τ-leaping, simulate and plot a single trajectory of the model

$$S \xrightarrow{1} 2S,$$

with $h = 0.01$, $X(0) = 100$, and a terminal time of $T = 10$. Now simulate and plot the same model using an exact simulation method. Compare the time it takes to generate each sample path.

Erratum to

Models of biochemical reaction systems

David F. Anderson and Thomas G. Kurtz

Department of Mathematics, University of Wisconsin, Madison, WI, USA
Departments of Mathematics and Statistics, University of Wisconsin, Madison, WI, USA

© Springer International Publishing Switzerland 2015
D.F. Anderson, T.G. Kurtz, *Stochastic Analysis of Biochemical Systems*,
Mathematical Biosciences Institute Lecture Series 1,
DOI 10.1007/978-3-319-16895-1

DOI 10.1007/978-3-319-16895-1_6

In the original publication, the Electronic Supplementary Material (ESM) was omitted in chapter 2. The updated ESM have now been added to Springer Link.

The online version of the updated original chapter can be found at
http://dx.doi.org/10.1007/978-3-319-16895-1_2

© Springer International Publishing Switzerland 2016
D.F. Anderson, T.G. Kurtz, *Stochastic Analysis of Biochemical Systems*,
Mathematical Biosciences Institute Lecture Series 1,
DOI 10.1007/978-3-319-16895-1_6

Appendix A
Notes on probability theory and stochastic processes

A.1 Some notation and basic concepts

A.1.1 Measure theoretic foundations of probability

Details of the following basic material on the measure theoretic foundations of probability can be found in any graduate text on the subject, in particular, [18].

A *measurable space* $(E, \mathcal{E})$ consists of a set E and a σ-algebra $\mathcal{E}$ of subsets of E. If E is a metric space, then, typically, $\mathcal{E}$ is taken to be $\mathcal{B}(E)$, the Borel subsets of E.

Let $(E_i, \mathcal{E}_i)$ be measurable spaces. A function $f : E_1 \to E_2$ is *measurable* if $f^{-1}(A) = \{x \in E_1 : f(x) \in A\} \in \mathcal{E}_1$ for each $A \in \mathcal{E}_2$.

Lemma A.1. *If $f : E_1 \to E_2$ and $g : E_2 \to E_3$ are measurable, then $g \circ f : E_1 \to E_3$ is measurable.*

A *probability space* $(\Omega, \mathcal{F}, P)$ is a model of an experiment in which Ω is the set of *possible outcomes*, $\mathcal{F}$ is the collection of *events,* where by an event A we mean the subset of Ω for which some "statement" is true ($\mathcal{F}$ is always taken to be a σ-algebra so that $(\Omega, \mathcal{F})$ is a measurable space), and P is a probability measure on $\mathcal{F}$, that is a measure with $P(\Omega) = 1$. $\mathcal{F}$ should contain all the events that correspond to statements in which the experimenter is interested, so in a sense, $\mathcal{F}$ represents the information that could be obtained by observing the complete experiment. With this interpretation of events, a sub-σ-algebra $\mathcal{D} \subset \mathcal{F}$ represents partial information, that is the information that would be obtained by checking the validity of only a subset of the statements.

A *random variable* X is some quantity that can be observed by performing the experiment. X might be real-valued, but in general can be vector-valued or even take values in a more general space E. Of course, the quantity depends on the outcome of the experiment, so X is a function defined on Ω and taking values

© Springer International Publishing Switzerland 2015
D.F. Anderson, T.G. Kurtz, *Stochastic Analysis of Biochemical Systems*,
Mathematical Biosciences Institute Lecture Series 1,
DOI 10.1007/978-3-319-16895-1

in E. We always assume that there is a σ-algebra $\mathscr{E}$ of subsets of E specified (so $(E,\mathscr{E})$ is a *measurable space*), and to say that X is a quantity that can be "observed" means that subsets of the form $\{X \in C\} = \{\omega \in \Omega : X(\omega) \in C\}$ are events in $\mathscr{F}$ for all $C \in \mathscr{E}$. In other words, X is a *measurable function* from $(\Omega,\mathscr{F})$ into $(E,\mathscr{E})$.

Mathematically, the *expectation* $E[X]$ of a real-valued random variable X is just the integral $\int_{\Omega} X(\omega)P(d\omega)$, assuming that the integral exists, but for the "real" meaning of expectation in probability theory, see the statement of the law of large numbers in any probability text or in Section 5.1.

Assuming that X is real-valued, i.e. that $E = \mathbb{R}$, and that $\mathscr{D}$ is a sub-σ-algebra of $\mathscr{F}$, a *conditional expectation* $E[X|\mathscr{D}]$ is the best estimate of X using the information in $\mathscr{D}$. If $E[X^2] < \infty$, then by "best estimate" using the information in $\mathscr{D}$, we mean the $\mathscr{D}$-measurable random variable Y (a random variable whose value is determined by the information in $\mathscr{D}$) that minimizes $E[(X - Y)^2]$. The condition $E[X^2] < \infty$ is not necessary and $E[X|\mathscr{D}]$ can be defined for all X satisfying $E[|X|] < \infty$. In general, if $E[|X|] < \infty$, then $E[X|\mathscr{D}]$ is the $\mathscr{D}$-measurable random variable Z satisfying

$$E[X\mathbf{1}_D] = E[Z\mathbf{1}_D] \quad \forall D \in \mathscr{D}.$$

If the information in $\mathscr{D}$ is the information obtained by observing random variables Y and Z (denoted $\mathscr{D} = \sigma(Y,Z)$, the smallest σ-algebra with respect to which Y and Z are measurable), then there is a function f_X such that

$$E[X|\sigma(Y,Z)] = f_X(Y,Z).$$

If $E[X^2] < \infty$, then f_X is the minimizer of $E[(X - f(Y,Z))^2]$,

$$E[(X - f_X(Y,Z))^2] = \inf_f E[(X - f(Y,Z))^2].$$

(Of course, f_X has to be appropriately measurable.)

A.1.2 Dominated convergence theorem

Theorem A.2. *Let $X_n \to X$ and $Y_n \to Y$ in probability. Suppose that $|X_n| \le Y_n$ a.s. and $E[Y_n|\mathscr{D}] \to E[Y|\mathscr{D}]$ in probability. Then*

$$E[X_n|\mathscr{D}] \to E[X|\mathscr{D}] \qquad \text{in probability.}$$

Proof. A sequence converges in probability iff every subsequence has a further subsequence that converges a.s., so we assume $Y_n \to Y$ and $E[Y_n|\mathscr{D}] \to E[Y|\mathscr{D}]$ almost surely. Let $D_{m,c} = \{\sup_{n \ge m} E[Y_n|\mathscr{D}] \le c\}$. Then

$$E[Y_n\mathbf{1}_{D_{m,c}}|\mathscr{D}] = E[Y_n|\mathscr{D}]\mathbf{1}_{D_{m,c}} \overset{L_1}{\to} E[Y|\mathscr{D}]\mathbf{1}_{D_{m,c}} = E[Y\mathbf{1}_{D_{m,c}}|\mathscr{D}].$$

Consequently, $E[Y_n 1_{D_{m,c}}] \to E[Y 1_{D_{m,c}}]$, so $Y_n 1_{D_{m,c}} \to Y 1_{D_{m,c}}$ in L_1 by the ordinary dominated convergence theorem. It follows that $X_n 1_{D_{m,c}} \to X 1_{D_{m,c}}$ in L_1 and hence

$$E[X_n | \mathscr{D}] 1_{D_{m,c}} = E[X_n 1_{D_{m,c}} | \mathscr{D}] \overset{L_1}{\to} E[X 1_{D_{m,c}} | \mathscr{D}] = E[X | \mathscr{D}] 1_{D_{m,c}}.$$

Since m and c are arbitrary, the lemma follows. □

A.1.3 General theory of stochastic processes

The following material can be found, for example, in [42] and Chapter 2 of [19].

In the study of stochastic processes (for our purposes, random functions of time $t \in [0, \infty)$), it is natural to consider the information that is available to an observer who has followed the experiment up to some time $t \geq 0$. That information is modeled by a sub-σ-algebra $\mathscr{F}_t \subset \mathscr{F}$. Assuming that the observer is not forgetful, $t < s$ implies $\mathscr{F}_t \subset \mathscr{F}_s$. The collection $\{\mathscr{F}_t\} = \{\mathscr{F}_t, t \geq 0\}$ of these σ-algebras is called a *filtration* (or sometimes a *history*). Since different observers may observe different aspects of the experiment, more than one filtration may be important in the model.

A stochastic process X is $\{\mathscr{F}_t\}$-*adapted* if for each $t \geq 0$, $X(t)$ is $\mathscr{F}_t$-measurable, that is, the value of $X(t)$ is part of the information known at time t.

A nonnegative random variable τ is a $\{\mathscr{F}_t\}$-*stopping time*, if for each $t \geq 0$, $\{\tau \leq t\} \in \mathscr{F}_t$. In other words, an observer whose information is modeled by $\{\mathscr{F}_t\}$ "sees" τ when it occurs. Intuitively, it makes sense to talk about the information $\mathscr{F}_\tau$ available to the observer at the stopping time τ. The formal mathematical definition is

$$\mathscr{F}_\tau = \{A \in \mathscr{F} : A \cap \{\tau \leq t\} \in \mathscr{F}_t, \forall t \geq 0\}.$$

Hitting times give a natural class of stopping times. If X is a cadlag (that is, is right continuous and has left limits at all $t > 0$) and adapted process and K is a closed subset, then

$$\tau_K^h = \inf\{t : X(t) \text{ or } X(t-) \in K\}$$

is a stopping time.

Note that a constant t is a stopping time, and if τ_1 and τ_2 are stopping times, then $\tau_1 \wedge \tau_2$ and $\tau_1 \vee \tau_2$ are stopping times. In particular $\tau_1 \wedge t$ is a stopping time.

A.2 Martingales

The material in this section can be found in Chapter 2 of [19].

An $\mathbb{R}$-valued stochastic process M adapted to $\{\mathscr{F}_t\}$ is an $\{\mathscr{F}_t\}$-*martingale* if $E[|M(t)|] < \infty$ for all $t \geq 0$ and

$$E[M(t+r) | \mathscr{F}_t] = M(t), \quad t, r \geq 0,$$

or equivalently,

$$E[M(t+r) - M(t)|\mathscr{F}_t] = 0.$$

An $\mathbb{R}$-valued stochastic process X adapted to $\{\mathscr{F}_t\}$ is a *submartingale* if $E[|X(t)|] < \infty$ for all $t \geq 0$ and

$$E[X(t+s)|\mathscr{F}_t] \geq X(t), \quad t, s \geq 0.$$

If X is a submartingale and τ_1 and τ_2 are stopping times, then the *optional sampling theorem* states

$$E[X(t \wedge \tau_2)|\mathscr{F}_{\tau_1}] \geq X(t \wedge \tau_1 \wedge \tau_2).$$

If τ_2 is finite a.s., $E[|X(\tau_2)|] < \infty$ and $\lim_{t \to \infty} E[|X(t)|\mathbf{1}_{\{\tau_2 > t\}}] = 0$, then

$$E[X(\tau_2)|\mathscr{F}_{\tau_1}] \geq X(\tau_1 \wedge \tau_2).$$

Of course, if X is a martingale

$$E[X(t \wedge \tau_2)|\mathscr{F}_{\tau_1}] = X(t \wedge \tau_1 \wedge \tau_2).$$

It follows that if M is a martingale and τ is a stopping time, then M^τ defined by

$$M^\tau(t) = M(\tau \wedge t)$$

is a martingale.

M is a *local martingale* if there exists a sequence of stopping times $\tau_n \to \infty$ such that M^{τ_n} is a martingale. We call the sequence $\{\tau_n\}$ a *localizing sequence* for M. Note that if

$$\lim_{n \to \infty} E[|M(t) - M(t \wedge \tau_n)|] = 0$$

for each $t \geq 0$, then the local martingale is, in fact, a martingale.

A.2.1 Doob's inequalities

Let X be a submartingale. Then for $x > 0$,

$$P\{\sup_{s \leq t} X(s) \geq x\} \leq x^{-1} E[X(t)^+]$$

$$P\{\inf_{s \leq t} X(s) \leq -x\} \leq x^{-1}(E[X(t)^+] - E[X(0)]).$$

If X is nonnegative and $\alpha > 1$, then

$$E[\sup_{s \leq t} X(s)^\alpha] \leq \left(\frac{\alpha}{\alpha - 1}\right)^\alpha E[X(t)^\alpha].$$

Note that by Jensen's inequality, if M is a martingale, then $|M|$ is a submartingale. In particular, if M is a square integrable martingale, then

$$E[\sup_{s \le t} |M(s)|^2] \le 4E[M(t)^2].$$

A.3 Stochastic integrals

For a full discussion of stochastic integration, see [42].

All of the stochastic processes we consider are *cadlag*, that is, have sample paths that are right continuous and have left limits at each $t > 0$. If X and Y are cadlag stochastic processes, then, if the following limit exists in probability, we can define the stochastic integral

$$\int_0^t X(s-)dY(s) \equiv \lim \sum_{i=1}^m X(s_{i-1})(Y(s_i) - Y(s_{i-1})), \qquad (A.1)$$

where $s_0 < s_1 < \cdots < s_m$ is a partition of $[0,t]$ and the limit is through any sequence of partitions with $\max\{s_{i+1} - s_i\} \to 0$.

We frequently consider integrals against counting processes, R,

$$\int_0^t X(s-)dR(s).$$

Note that this integral is not necessarily a Stieltjes integral. For example,

$$\int_0^t R(s-)dR(s) = \frac{R(t)(R(t) - 1)}{2},$$

by (A.1), but $\int R\,dR$ does not exist as a Stieltjes integral because the integrand and integrator have common discontinuities. In particular,

$$\lim \sum R(s_i)(R(s_i) - R(s_{i-1})) = \frac{(R(t) + 1)R(t)}{2},$$

where the limit is taken as in (A.1).

We also note that if M is a (local) martingale with respect to a filtration $\{\mathscr{F}_t\}$ and X is a cadlag, $\{\mathscr{F}_t\}$-adapted process, then

$$Z(t) = \int_0^t X(s)dM(s)$$

is at least a local martingale. If $\{\tau_n\}$ is a localizing sequence for M and $\sigma_n = \inf\{t : |X(t)| \vee |X(t-)| \ge n\}$, then $\{\tau_n \wedge \sigma_n\}$ is a localizing sequence for Z.

A.4 Convergence in distribution and the functional central limit theorem for Poisson processes

All of the material in this section can be found in [11] or Chapter 3 of [19].

Let (S,d) be a complete, separable metric space, and let $C_b(S)$ denote the space of bounded, continuous functions on S. We say that a sequence of S-valued random variables $\{X_n\}$ *converges in distribution* to X (or equivalently, $\{P_{X_n}\}$ *converges weakly* to P_X, where P_{X_n} and P_X denote the distributions of X_n and X) if for each $f \in C_b(S)$

$$\lim_{n \to \infty} E[f(X_n)] = E[f(X)].$$

We will denote convergence in distribution by $X_n \Rightarrow X$.

There are many equivalent ways of specifying convergence in distribution. For example, we have the following.

Lemma A.3. $\{X_n\}$ *converges in distribution to* X *if and only if*

$$\liminf_{n \to \infty} P\{X_n \in A\} \geq P\{X \in A\}, \text{ each open } A,$$

or equivalently

$$\limsup_{n \to \infty} P\{X_n \in B\} \leq P\{X \in B\}, \text{ each closed } B.$$

The simplest way to understand the implications of convergence in distribution is through the Skorohod representation theorem.

Theorem A.4. *Suppose that* $X_n \Rightarrow X$. *Then there exists a probability space* $(\Omega, \mathscr{F}, P)$ *and random variables,* $\tilde{X}_n$ *and* $\tilde{X}$, *such that* $\tilde{X}_n$ *has the same distribution as* X_n, $\tilde{X}$ *has the same distribution as* X, *and* $\tilde{X}_n \to \tilde{X}$ *a.s.*

The following corollary is an immediate consequence of the Skorohod representation theorem.

Corollary A.5. *Let* (S,d) *and* (E,r) *be complete, separable metric spaces, and let* $G : S \to E$ *be a Borel measurable function. Define*

$$C_G = \{x \in S : G \text{ is continuous at } x\}.$$

Suppose $X_n \Rightarrow X$ *and that* $P\{X \in C_G\} = 1$. *Then* $G(X_n) \Rightarrow G(X)$.

In the study of stochastic processes, S is typically a function space, and the most important function space is the space of cadlag functions $x : [0,\infty) \to E$, where (E,r) is a complete, separable metric space and by cadlag we mean the function x is right continuous and has left limits at each $t > 0$. We denote this space by $D_E[0,\infty)$. Skorohod defined a topology on $D_E[0,\infty)$ by requiring that $x_n \to x$ if and only if there exists a sequence of strictly increasing functions λ_n from $[0,\infty)$ onto $[0,\infty)$ such that for each $t > 0$,

$$\lim_{n\to\infty}\sup_{s\leq t}|\lambda_n(s)-s|=0 \quad\text{and}\quad \lim_{n\to\infty}\sup_{s\leq t}r(x_n\circ\lambda_n(s),x(s))=0.$$

A somewhat subtle fact is that this topology corresponds to a metric d under which $D_E[0,\infty)$ is a complete separable metric space. For our purposes, it is enough to know that if x is continuous, then convergence of x_n to x is equivalent to

$$\lim_{n\to\infty}\sup_{s\leq t}r(x_n(s),x(s))=0 \quad\text{for all}\quad t>0,$$

since in the following lemma, the limit is continuous.

Lemma A.6. *Let Y be a unit Poisson process and $\tilde{Y}(u)=Y(u)-u$. Define*

$$W^N(u)=\frac{1}{\sqrt{N}}\tilde{Y}(Nu).$$

Then $W^N\Rightarrow W$, where W is a standard Brownian motion.

Proof. The result follows from the classical central limit theorem once one proves relative compactness of the sequence. It can also be obtained as an immediate corollary of the martingale functional central limit theorem, for example, Theorem 7.1.4 of [19]. $\qquad\square$

In applying this lemma, by the Skorohod representation theorem, we can pretend that for each $u_0>0$,

$$\sup_{u\leq u_0}|W^N(u)-W(u)|\to 0,$$

but one should note that

$$(W^N,W^{N^2})\Rightarrow(W_1,W_2),$$

where W_1 and W_2 are independent standard Brownian motions.

A.5 Conditioning and independence

Let $(\Omega,\mathscr{F},P)$ be a probability space, and let $\mathscr{A},\mathscr{B}\subset\mathscr{F}$ be sub-σ-algebras. Then $\mathscr{A}$ and $\mathscr{B}$ are *independent* if $P(A\cap B)=P(A)P(B)$ for all $A\in\mathscr{A}$ and $B\in\mathscr{B}$. A random variable X is independent of $\mathscr{A}$ if $\sigma(X)$ (the smallest σ-algebra with respect to which X is measurable) is independent of $\mathscr{A}$.

Let $(E,\mathscr{E})$ be a measurable space. A collection of bounded, measurable functions $\mathscr{S}$ is *separating* for finite measures on $\mathscr{E}$ if for finite measures μ,ν, $\int_E f d\mu=\int_E f d\nu$ for all $f\in\mathscr{S}$ implies $\mu=\nu$. For example, for $E=\mathbb{R}$, $\mathscr{S}=\{f(x)=e^{i\theta x}:\theta\in\mathbb{R}\}$ is separating.

If (E,r) is a complete, separable metric space, then $B(E)$ denotes the space of bounded, Borel measurable functions.

Lemma A.7. *Let (E,r) be a complete, separable metric space and $\mathscr{S} \subset B(E)$ be separating. Let X be an E-valued random variable and $\mathscr{D} \subset \mathscr{F}$ be a sub-σ-algebra. Suppose that for each $f \in \mathscr{S}$, $E[f(X)|\mathscr{D}] = E[f(X)]$. Then X is independent of $\mathscr{D}$.*

Proof. Let $D \in \mathscr{D}$, and for $A \in \mathscr{B}(E)$, define

$$\mu_D(A) = E[\mathbf{1}_A(X)\mathbf{1}_D] \text{ and } \nu_D(A) = E[\mathbf{1}_A(X)]P(D).$$

For $f \in \mathscr{S}$,

$$\int f d\mu_D = E[f(X)\mathbf{1}_D] = E[f(X)]P(D) = \int f d\nu_D,$$

which, since $\mathscr{S}$ is separating, implies $\mu_D = \nu_D$. Since $D \in \mathscr{D}$ is arbitrary, we have

$$P(\{X \in A\} \cap D) = \mu_D(A) = \nu_D(A) = P\{X \in A\}P(D)$$

for all $A \in \mathscr{B}(E)$ and $D \in \mathscr{D}$. Consequently, X and $\mathscr{D}$ are independent.   □

A.6 Directed sets

A directed set is a nonempty set $\mathscr{I}$ together with a reflexive and transitive binary relation $\preceq$, such that every pair of elements has an upper bound. That is, for any a and b in $\mathscr{I}$ there must exist a c in $\mathscr{I}$ with $a \preceq c$ and $b \preceq c$.

A.7 Gronwall inequality

Lemma A.8. *Suppose that A is nonnegative, cadlag, and non-decreasing. Further suppose that X is cadlag, and that*

$$0 \le X(t) \le \varepsilon + \int_0^t X(s-)dA(s) . \tag{A.2}$$

Then

$$X(t) \le \varepsilon e^{A(t)}.$$

Proof. Iterating (A.2),

$$X(t) \le \varepsilon + \int_0^t X(s-)dA(s)$$

$$\le \varepsilon + \varepsilon A(t) + \int_0^t \int_0^{s-} X(u-)dA(u)dA(s)$$

$$\le \varepsilon + \varepsilon A(t) + \varepsilon \int_0^t A(s-)dA(s) + \int_0^t \int_0^{s-} \int_0^{u-} X(r-)dA(r)dA(u)dA(s).$$

Since A is finite variation, making $[A]^c_t \equiv 0$, Itô's formula yields

$$e^{A(t)} = 1 + \int_0^t e^{A(s-)} dA(s) + \Sigma_{s \leq t}(e^{A(s)} - e^{A(s-)} - e^{A(s-)}\Delta A(s))$$

$$\geq 1 + \int_0^t e^{A(s-)} dA(s)$$

$$\geq 1 + A(t) + \int_0^t \int_0^{s-} e^{A(u-)} dA(u) dA(s)$$

$$\geq 1 + A(t) + \int_0^t A(s-) dA(s) + \int_0^t \int_0^{s-} \int_0^{u-} e^{A(r-)} dA(r) dA(u) dA(s).$$

Continuing the iteration, we see that $X(t) \leq \varepsilon e^{A(t)}$. $\qquad\qquad \square$

References

[1] David F. Anderson. A modified next reaction method for simulating chemical systems with time dependent propensities and delays. *J. Chem. Phys.*, 127(21): 214107, 2007.

[2] David F. Anderson. Incorporating postleap checks in tau-leaping. *J. Chem. Phys.*, 128(5):054103, 2008.

[3] David F. Anderson. A proof of the Global Attractor Conjecture in the single linkage class case. *SIAM J. Appl. Math*, 71(4):1487–1508, 2011.

[4] David F. Anderson and Desmond J. Higham. Multi-level Monte Carlo for continuous time Markov chains, with applications in biochemical kinetics. *SIAM: Multiscale Modeling and Simulation*, 10(1):146–179, 2012.

[5] David F. Anderson and Masanori Koyama. Weak error analysis of numerical methods for stochastic models of population processes. *SIAM: Multiscale Modeling and Simulation*, 10(4):1493–1524, 2012.

[6] David F. Anderson, Gheorghe Craciun, and Thomas G. Kurtz. Product-form stationary distributions for deficiency zero chemical reaction networks. *Bull. Math. Biol.*, 72(8):1947–1970, 2010.

[7] David F. Anderson, Arnab Ganguly, and Thomas G. Kurtz. Error analysis of tau-leap simulation methods. *Annals of Applied Probability*, 21(6):2226–2262, 2011.

[8] David F. Anderson, Germán A. Enciso, and Matthew D. Johnston. Stochastic analysis of biochemical reaction networks with absolute concentration robustness. *J. R. Soc. Interface*, 11(93):20130943, 2014a.

[9] David F. Anderson, Desmond J. Higham,, and Yu Sun. Complexity analysis of multilevel Monte Carlo tau-leaping. accepted to SIAM Journal on Numerical Analysis, 2014b.

[10] Karen Ball, Thomas G. Kurtz, Lea Popovic, and Greg Rempala. Asymptotic analysis of multiscale approximations to reaction networks. *Ann. Appl. Probab.*, 16(4):1925–1961, 2006. ISSN 1050-5164.

© Springer International Publishing Switzerland 2015
D.F. Anderson, T.G. Kurtz, *Stochastic Analysis of Biochemical Systems*,
Mathematical Biosciences Institute Lecture Series 1,
DOI 10.1007/978-3-319-16895-1

[11] Patrick Billingsley. *Convergence of probability measures*. Wiley Series in Probability and Statistics: Probability and Statistics. John Wiley & Sons Inc., New York, second edition, 1999. ISBN 0-471-19745-9. doi 10.1002/ 9780470316962. URL http://dx.doi.org/10.1002/9780470316962. A Wiley-Interscience Publication.

[12] K. Burrage and T. Tian. Poisson Runge-Kutta methods for chemical reaction systems. In Y. Lu, W. Sun, and T. Tang, editors, *Advances in Scientific Computing and Applications*, pages 82–96. Science Press, 2003.

[13] Yang Cao, Daniel T. Gillespie, and Linda R. Petzold. Avoiding negative populations in explicit Poisson tau-leaping. *J. Chem. Phys.*, 123:054104, 2005.

[14] Yang Cao, Daniel T. Gillespie, and Linda R. Petzold. Efficient step size selection for the tau-leaping simulation method. *J. Chem. Phys.*, 124:044109, 2006.

[15] Abhijit Chatterjee and Dionisios G. Vlachos. Binomial distribution based τ-leap accelerated stochastic simulation. *J. Chem. Phys.*, 122:024112, 2005.

[16] Gheorghe Craciun, Alicia Dickenstein, Anne Shiu, and Bernd Sturmfels. Toric dynamical systems. *J. Symbolic Comput.*, 44(11):1551–1565, 2009.

[17] Thomas A. Darden. Enzyme kinetics: stochastic vs. deterministic models. In *Instabilities, bifurcations, and fluctuations in chemical systems (Austin, Tex., 1980)*, pages 248–272. Univ. Texas Press, Austin, TX, 1982.

[18] Rick Durrett. *Probability: theory and examples*. Cambridge Series in Statistical and Probabilistic Mathematics. Cambridge University Press, Cambridge, fourth edition, 2010. ISBN 978-0-521-76539-8. doi 10. 1017/CBO9780511779398. URL http://dx.doi.org/10.1017/ CBO9780511779398.

[19] Stewart N. Ethier and Thomas G. Kurtz. *Markov Processes: Characterization and Convergence*. John Wiley & Sons, New York, 1986.

[20] M. Feinberg. Lectures on chemical reaction networks. Delivered at the Mathematics Research Center, Univ. Wisc.-Madison. Available for download at http://crnt.engineering.osu.edu/LecturesOnReactionNetworks, 1979.

[21] M. Feinberg. Chemical reaction network structure and the stability of complex isothermal reactors - I. the deficiency zero and deficiency one theorems, review article 25. *Chem. Eng. Sci.*, 42:2229–2268, 1987.

[22] Chetan Gadgil, Chang Hyeong Lee, and Hans G. Othmer. A stochastic analysis of first-order reaction networks. *Bull. Math. Biol.*, 67(5):901–946, 2005. ISSN 0092-8240. doi 10.1016/j.bulm.2004.09.009. URL http://dx.doi. org/10.1016/j.bulm.2004.09.009.

[23] M.A. Gibson and J. Bruck. Efficient exact stochastic simulation of chemical systems with many species and many channels. *J. Phys. Chem. A*, 105:1876–1889, 2000.

[24] Mike B. Giles. Multilevel Monte Carlo path simulation. *Operations Research*, 56:607–617, 2008.

[25] D. T. Gillespie. A general method for numerically simulating the stochastic time evolution of coupled chemical reactions. *J. Comput. Phys.*, 22:403–434, 1976.

[26] D. T. Gillespie. Exact stochastic simulation of coupled chemical reactions. *J. Phys. Chem.*, 81(25):2340–2361, 1977.

[27] D. T. Gillespie and Linda R. Petzold. Improved leap-size selection for accelerated stochastic simulation. *J. Chem. Phys.*, 119(16):8229–8234, 2003.

[28] Daniel T. Gillespie. Approximate accelerated stochastic simulation of chemically reacting systems. *J. Chem. Phys.*, 115(4):1716–1733, 2001. doi http://dx.doi.org/10.1063/1.1378322. URL http://scitation.aip.org/content/aip/journal/jcp/115/4/10.1063/1.1378322.

[29] J. Gunawardena. Chemical reaction network theory for in-silico biologists. Notes available for download at http://vcp.med.harvard.edu/papers/crnt.pdf, 2003.

[30] Eric L. Haseltine and James B. Rawlings. Approximate simulation of coupled fast and slow reactions for stochastic chemical kinetics. *J. Chem. Phys.*, 117(15):6959–6969, 2002.

[31] Stefan Heinrich. Multilevel Monte Carlo methods. *Springer, Lect. Notes Comput. Sci.*, 2179:58–67, 2001.

[32] F. J. M. Horn and R. Jackson. General mass action kinetics. *Arch. Rat. Mech. Anal.*, 47:81–116, 1972.

[33] Jean Jacod. Multivariate point processes: predictable projection, Radon-Nikodým derivatives, representation of martingales. *Z. Wahrscheinlichkeitstheorie und Verw. Gebiete*, 31:235–253, 1974/75.

[34] Hye-Won Kang and Thomas G. Kurtz. Separation of time-scales and model reduction for stochastic reaction networks. *Ann. Appl. Probab.*, 23(2):529–583, 2013. doi 10.1214/12-AAP841. URL http://projecteuclid.org/euclid.aoap/1360682022.

[35] Frank P. Kelly. *Reversibility and Stochastic Networks*. John Wiley & Sons Ltd., Chichester, 1979. ISBN 0-471-27601-4. Wiley Series in Probability and Mathematical Statistics.

[36] Thomas G. Kurtz. The optional sampling theorem for martingales indexed by directed sets. *Ann. Probab.*, 8(4):675–681, 1980a. ISSN 0091-1798. URL http://links.jstor.org/sici?sici=0091-1798(198008)8:4<675:TOSTFM>2.0.CO;2-7&origin=MSN.

[37] Thomas G. Kurtz. Representations of Markov processes as multiparameter time changes. *Ann. Probab.*, 8(4):682–715, 1980b. ISSN 0091-1798. URL http://links.jstor.org/sici?sici=0091-1798(198008)8:4<682:ROMPAM>2.0.CO;2-W&origin=MSN.

[38] Tiejun Li. Analysis of explicit tau-leaping schemes for simulating chemically reacting systems. *SIAM Multiscale Model. Simul.*, 6(2):417–436, 2007.

[39] Arjun Kumar Manrai and Jeremy Gunawardena. The geometry of multisite phosphorylation. *Biophys. J.*, 95(12), 2008.

[40] P. A. Meyer. Démonstration simplifiée d'un théorème de Knight. In *Séminaire de Probabilités, V (Univ. Strasbourg, année universitaire 1969–1970)*, pages 191–195. Lecture Notes in Math., Vol. 191. Springer, Berlin, 1971.

[41] M.P. Millán, A. Dickenstein, A. Shiu, and C. Conradi. Chemical reaction systems with toric steady states. *Bull. Math. Biol.*, 74(5):1027–1065, 2012.

[42] Philip E. Protter. *Stochastic integration and differential equations*, volume 21 of *Stochastic Modelling and Applied Probability*. Springer-Verlag, Berlin, 2005. ISBN 3-540-00313-4. doi 10.1007/978-3-662-10061-5. URL http://dx.doi.org/10.1007/978-3-662-10061-5. Second edition. Version 2.1, Corrected third printing.

[43] Muruhan Rathinam, Linda R. Petzold, Yang Cao, and Daniel T. Gillespie. Stiffness in stochastic chemically reacting systems: The implicit tau-leaping method. *J. Chem. Phys.*, 119:12784–12794, 2003.

[44] Muruhan Rathinam, Linda R. Petzold, Yang Cao, and Daniel T. Gillespie. Consistency and stability of tau-leaping schemes for chemical reaction systems. *SIAM Multiscale Model. Simul.*, 3:867–895, 2005.

[45] R. Srivastava, L. You, J. Summers, and J. Yin. Stochastic vs. deterministic modeling of intracellular viral kinetics. *J. Theoret. Biol.*, 218(3):309–321, 2002. ISSN 0022-5193.

[46] T. Tian and K. Burrage. Binomial leap methods for simulating stochastic chemical kinetics. *J. Chem. Phys.*, 121:10356, 2004.

[47] Shinzo Watanabe. On discontinuous additive functionals and Lévy measures of a Markov process. *Japan. J. Math.*, 34:53–70, 1964.

Index

© Springer International Publishing Switzerland 2015
D.F. Anderson, T.G. Kurtz, *Stochastic Analysis of Biochemical Systems*,
Mathematical Biosciences Institute Lecture Series 1,
DOI 10.1007/978-3-319-16895-1